BIRDS IN HAND

RCA And A Communication Revolution

ATM Consulting

BIRDS IN HAND - RCA AND A COMMUNICATION REVOLUTION
by Archie T. Miller

Front Cover Illustration - **Satcom-1 On Station** - *RCA Astro Electronics Division*
Back Cover Photo - **Satcom-2R Launch** - *NASA*

ISBN-13: 978-1466219878

ISBN-10: 1466219874

Table of Contents

FOREWORD

This book was written about 35 years after the RCA SATCOM saga began.   Many of those intimately involved are now dead -- I should have started this project sooner.   For those still around, recollections are sometimes biased by the ravages of time.   Records are harder to find and access now.   As much as I could, I have tried to verify facts and find multiple sources, however, this book is a story not news reporting.   My primary purpose is to tell the story of a communications technology and a business that forever changed the way we all receive, news, entertainment, sports and many other kinds of data.   The truth is that if there had not been communication satellites we would not have cable television and digital communications as we know it today.   The undeniable fact is that RCA led the field in this endeavor, utilizing leading edge technology and creative engineering, and made a lot of money in the process.   We also proved that with a trained, disciplined, operations team you can provide reliable high quality communications despite severe technical problems in complex systems 23,000 miles away in space.

RCA as a corporate entity is long gone.   The satellite communication enterprise that started in the early 1970's as RCA Glōbcom became RCA Americom in 1975.   In 1986 when GE bought RCA it became GE Americom.   GE sold Americom to Société Européenne des Satellites (SES) in 2001 and it then became SES Americom.   In 2010 Americom was folded into SES Engineering and operationally became SES World Skies -- RIP Americom.

Those of us who were involved in this revolution could never have predicted which way it would go and where it would end. In about 1971 someone at Comsat wrote a paper about the requirements for a US Domestic Satellite System. It was stated that the requirements for satellite TV capacity would be driven by coverage of the National Football League games on a Sunday afternoon -- about 7 transponders!   By 1982, Americom alone had more than 50 transponders carrying TV.   In 1975 when RCA launched its first satellite, the primary business opportunity was considered to be Private Leased Channel telephone communications -- 10 years later that business was wiped out by the breakup of AT&T, technical constraints and fiber optics. The marriage of the single-point to multi-point capability of communications satellites and the struggling mom-and-pop Community Antenna TV Systems spawned the cable-centered multi-media world we know today. Who knew?

Those were exciting and interesting times. Leading Americom's Spacecraft Operations for 18-years was the best job I ever had.   And it was certainly challenging -- the evolving spacecraft of those times required intervention from the ground to do almost everything.   And when things on the spacecraft broke, that intervention became even more complicated. By contrast, todays communication spacecraft are highly automated and require relatively little intervention from the ground.

 Now at 81, I thought that I would like to provide some record of an experience that, as Walter Cronkite used to say, "altered and illuminated our times".   I am certain that for the reasons noted above, I may have not given credit where it was due, misstated facts or erroneously described events -- if I did so, I am sorry.   I hope, however, that I have provided some insight into an important period in the history of communications.

I have tried to present this story chronologically.   In some cases, however, it is necessary to describe related events that may have occurred over a period of years for reasons of continuity and understanding.   You will have to excuse me if the chronology jumps back and forth a bit.

Archie T. Miller
Manchester, NJ
August 30, 2011

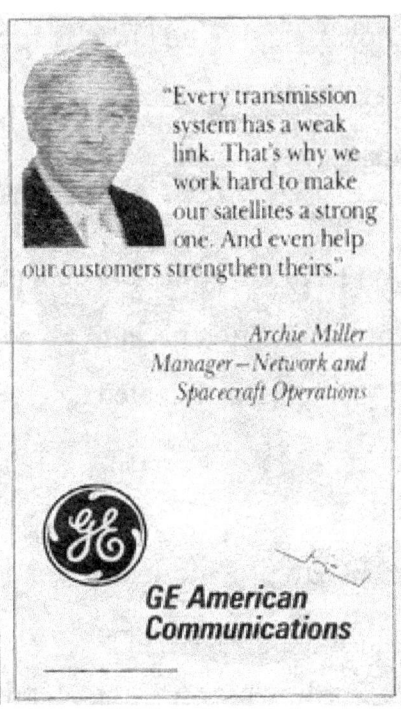
Multichannel News

ACKNOWLEDGEMENTS

Although much of this book is based on my own recollection, many people provided their memories, documents and recordings.

I am deeply indebted to Bob Youngblood, a colleague and long time friend for his anecdotes and technical knowledge but most of all for access to his treasure trove of memos, emails, reports and technical papers that he has maintained over all these years. Bob was with me from the beginning of this adventure in the early days at Princeton Junction. Fortunately his memory and records are better than mine and for that I am eternally grateful.

To Steve Agid, who fortunately for my failing memory, maintained a written journal throughout most of his adult life. He also has a large collection of notes, photos and memorabilia collected over many years.

To Bob Cooper, my sometime nemesis during our F-2 problems in the late '70s, now living in New Zealand, my thanks for his research through his files of "ladder charts" showing loading of the spacecraft during the '70s and '80s. Also for the CD's containing copies of the Community Antenna TV Journal and Coop's Satellite Digest and a DVD of his extensive TV interviews of all of us at Vernon Valley in 1978.

To my colleagues in Spacecraft Operations and the Vernon Valley Earth Station especially John Bailey, Mike Harrison, Ed Ingram, Joe Massey, John "Jack" Schmidt, T. P. Tubbs and Dave Whelan for their technical expertise and recollection of events long past.

To John Christopher for his many email answers to my questions on the technical and business details of the development of the Americom spacecraft programs.

To John Williamson for his recollection of the many public relations adventures we shared and for the cartoon of Nipper listening for the lost F-3.

To Carl Cangelosi, Murray Fruchter, Rick Langhans, Gerry Long and Bill Palme for their recollections of events in the early days of Americom.

To Bill Berman, Fred Caine, Jack Hyle and John Nelsen for their anecdotes and stories of the marketing and sales history of Americom. I am particularly indebted to Bill for his detailed story of the F-4 marketing strategy after the departure of HBO and how Crimson Partners got started.

To Bob Cenker, Steve Fox and George Martch of RCA Astro for their technical knowledge and help filling in details of the Satcom program.

To Leslie Gadsby of SES New Skies for her help in letting me go through the voluminous material collected for the Satcom-1 25th Anniversary celebration.

To Judy, my wife, for her patience and understanding during the long hours I have spent glued to the computer.

I must of course acknowledge the greatest source of all in this digital age -- the Internet. Wikipedia was an excellent starting point on many searches and provided leads for other searches. Google Books was an invaluable source for technical books, journals and papers. Undoubtedly one of the most valuable sources to this project was the Hauser Oral History Collection on the Cable Center Site. It contains in depth interviews with all the pioneers and major figures in the cable industry. Most helpful to me were the interviews with Ed Horowitz, Amos Hostetter, Kay Koplovitz, Brian Lamb, Gerald Levin, John Malone, Monty Rifkin, John Sie, Hub Schlafly, Sidney Topol and Ted Turner.

These sites are an excellent reference for technical data:
>
> *Gunter's Space Page* - space.skyrocket.de
> *Encyclopedia Astronautix* - astronautix.com
> *National Space Science Data Center* - nssdc.gsfc.nasa.gov

The New York Times Archive provided contemporaneous accounts of many of the important events in Americom's history.

These publications were helpful for background, dates and technical data.

Satellite Television Distribution - Andrew F. Inglis -
>
> Presented at the Royal Television Society London, England November 6, 1980

Trouble With Tanks AIAA 96-1147 - Robert F. Youngblood, Senior Member
>
> American Institute of Aeronautics and Astronautics.

Maximum Communication Capacity per Unit Cost - J. E. Keigler and C. R. Hume -
>
> RCA Astro Electronics Division

The Satellite Communication Application Handbook - Bruce R. Elbert -Artech House

The Wired Nation;: Cable TV - Ralph Lee Smith - Harper Colophon Books

Communication Satellites 4th Edition - Donald H. Martin - Aerospace Press

Satellite Technology: an Introduction - Andrew F. Inglis, Arch C. Luther - Focal Press

Satellite Circuit Magazine - A Publication of RCA Americom

C & ES News - A Publication of RCA Communications and Electronic Services

Archie T. Miller
Manchester, NJ

Steven J. Agid Photo

Chapter 1 - Lost In Space

In November of 1979, I was preparing to support the launch of RCA American Communications (Americom's) third communication satellite, Satcom 3 also called F-3. In 1975, when RCA decided to enter the communication satellite business, I became Americom's first Spacecraft Operations Manager. I was responsible for operating their new satellites built by RCA Astro-Electronics Division. Americom had successfully put two satellites into commercial service, one in December 1975 and one in March 1976. The business was going very well, both satellites were almost sold out and Americom needed more capacity. F-3 would provide that capacity and in fact almost one half of F-3's capacity had already been sold. I was looking forward to the F-3 launch and being part of an enterprise that was growing and had a bright future.

I had started my career with RCA with the RCA Service Co. Government Services Division in Alexandria, VA. in February 1957 as a Technical Writer. Within three years, I had advanced to Manager, Technical Publications responsible for a 50 person department providing technical documentation for the Army, Navy , Air Force and Coast Guard. In July 1965, I transferred to a new RCA Service Co. project at Goddard Space Flight Center (GSFC) as Manager, Technical Support. I was responsible for providing engineering and technical services for Goddard's worldwide satellite tracking network the Space Tracking And Data Acquisition Network (STADAN). In 1974 when Congress changed government procurement policy, RCA lost its major project at Goddard and seemed doomed to lose the others. As a result, I left RCA in June 1974 to join the American Satellite Corporation (ASC), a startup satellite communications company, as Manager, Network Operations. ASC was very disorganized and fraught with internal turf wars and I was beginning to think that I had made a big mistake. In less then a year, I worked for three different Directors reporting to three different VP's and I was beginning to suspect that my current boss was not being completely truthful with me. Fortunately, in June 1975, RCA asked me to rejoin the company as they prepared to launch their first communications satellites.

In December 1975 Americom had launched Satcom 1 (F-1) and in February 1976 Satcom 2 (F-2). These two satellites had been a great commercial success. F-1 had HBO as its anchor tenant and its remaining capacity was rapidly being sold out to other cable TV channels such as ESPN, WTBS Superstation, Showtime, USA Network and Christian Broadcast Network. F-2 was partially owned by RCA Alaska Communications (Alascom) and carried 14 transponders of telephone and television communications for Alaska including the innovative Bush Telephone Service. In addition it carried 7 transponders of Americom's Private Leased Channel (PLC) telephone service. By 1978, with F-1 almost completely filled with cable television traffic, Americom had started plans to launch Satcom 3. Americom's marketing department was actively working to sell out F-3 before it was launched. It wasn't too hard -- demand for satellite space from the booming cable TV business was greater than supply. Ted Turner whose WTBS Superstation had been a huge success on F-1 bought two channels on F-3; one for his proposed 24-hour news service, Cable News Network (CNN) and the other to move WTBS from F-1 to F-3.. From my standpoint, all of that was interesting, however, my prime responsibility was to keep F-1 and F-2 flying and prepare the Americom facilities and personnel to assist RCA Astro in the launch of F-3.

In the fall of 1979 as we were preparing to launch F-3, the United States was in turmoil. On November 4, Iranian militants had invaded and taken over the U.S. Embassy in Teheran, Iran. Sixty-six Americans were seized and held hostage by Islamist students and militants. The hostage-taking was the culmination of a long period of strained relations between the U.S. and Iran. In 1953 a CIA organized coup had removed the democratically-elected government of Mohammad Mosaddegh and replaced it with one headed by Shah Reza Pahlavi. In January 1979 Islamic fundamentalists removed the Shah from the throne and sent him into exile in Europe. In February, while Iran was being ruled by an "interim Islamic government", student and militants had briefly occupied the American Embassy. As a result of this incident, and recognizing the embassy's vulnerability, the embassy staff was reduced from about 1000 to 60 persons. PanAm operated 10 "Freedom Flights" from February 16 to 20 to get embassy personnel and other Americans out of Tehran. In October 1979, US President Carter decided

12

to allow the Shah of Iran to come to the U.S. for medical treatment despite advice to the contrary from the US Embassy staff in Teheran.   As predicted by the embassy staff, this further infuriated the Iranians because they considered the Shah a puppet of the U.S.   The Shah coming to the US for gall bladder surgery was probably the final straw that triggered the Teheran Embassy takeover by Iranian militants on November 4, 1979.   At about 6:30AM about 300 militants cut the chain on the gate and stormed the embassy.   The occupiers bound and blindfolded the staff and the marines and paraded them in front of photographers.

In the United States, the hostage-taking was seen as an outrage, violating a centuries-old principle of international law granting diplomats immunity from arrest and diplomatic compounds sovereignty in their embassies.   The Carter administration imposed economic sanctions on Iran and froze about $8-billion of Iranian assets in the U.S.   Naval Task Force 77.4 was moved into the Persian Gulf off the coast of Iran.   News headlines every day screamed the latest developments in the crisis.   TV news channels were almost completely dedicated to developments, rumors and commentary on the hostage crisis.

In the fall of 1979, F-1 and F-2 were on-station and operating at near capacity with revenue producing traffic.   As we went about our preparations for the F-3 launch, we were acutely aware of the crisis unfolding in Iran and its affect on the American people.   Television monitors of our F-1 and F-2 satellite TV traffic were filled with Iran news and other programming about the crisis all day long.

RCA Astro-Electronics in East Windsor, NJ had worked through 1979 completing construction and testing of RCA Satcom-3.   The Satcom project was on the leading edge of communication spacecraft technology at that time.   In the initial design of the Series 1000 Spacecraft for F-1, F-2 and F-3 Astro scientists knew that a cost-effective spacecraft for the competitive North American market would require a 24-channel spacecraft that could operate through all seasons and be able to provide services for at least 8 years.   They realized that such an innovative spacecraft would be too heavy for launch into geosynchronous orbit by existing launch rockets. RCA then embarked on a project to redesign the NASA Delta Launch Vehicle so that it could lift the heavier load (2000 lbs) of the new Satcom spacecraft. This was the first time that a commercial company had redesigned a NASA launch vehicle to increase its capability. The net result was that RCA had higher capacity satellites than their competitors and wanted to get them launched and sold out while they still had a technical competitive advantage.

On November 19th and 20th, as rehearsals and final preparations for the F-3 launch were reaching a peak, the Iranians released 13 of the hostages. This, of course, brought the crisis back to the front pages again and heightened speculation that all of the hostages might be released. This was not to be and the remaining hostages would be held for a total of 444 days. As the launch team worked toward the December 6 launch, all of the TV monitors displayed endless Iranian crisis stories carried by Americom's F-1 customers. And every day on the news broadcasts there was a countdown "Seventeenth day of the hostage crisis -- Eighteenth day of the hostage crisis -- "

During the first week in October, the Delta first stage and interstage were erected on Pad A at Launch Complex 17. The nine solid, strap-on rocket motors were mounted in place around the base of the first stage and the second stage was mated with the first. The F-3 Spacecraft underwent final testing at RCA Astro during October and was sealed in its shipping container (can) and shipped to Cape Canaveral in early November. On November 5, the F-3 spacecraft underwent its initial checkout in Hanger AE. There is a four week campaign in preparation for a Delta launch. This includes unpacking and checking out the spacecraft, exhaustive electric and mechanical testing, fueling the spacecraft with hydrazine fuel, performing spin balance tests and installing the spacecraft on top of the Delta 3914 launch vehicle. F-3 was moved to the Delta Spin Test Facility for final processing and mating with the Delta third stage on November 14. Launch is scheduled for Thursday of the fourth week of the campaign. This allows for three days (Mon - Wed) to fuel the Delta launch vehicle and do final tests and a backup day without weekend overtime (Friday) in case the launch is scrubbed on Thursday. The spacecraft, attached to the third stage, was moved to Complex 17 and erected atop the Delta. The payload fairing which will protect it during its flight through the atmosphere was emplaced. As it was, the launch was scrubbed on Thursday because of high winds aloft and Satcom 3 was finally launched perfectly late on Friday afternoon December 7, 1979.

The Delta Rocket placed F-3 in an orbit where the lowest point (perigee) was about 90 miles above the earth and the highest point (apogee) was about 22,000 miles above the earth. The plane of the orbit was inclined about 28 degrees to the equator. For the next two days there was a very high level of activity at the Americom Vernon Valley, NJ and South Mountain, CA stations and at the two Astro overseas stations in Carnarvon, Australia and Fucino, Italy. Range data was taken to refine the spacecraft orbit. Sun sensor and horizon sensor data was taken to determine the spacecraft's attitude (where it was pointing). Based on the collected attitude data, thrusters (small rockets) on F-3 were fired in Spin Precession Maneuvers to adjust the attitude. All of this was in preparation for firing the Apogee Kick Motor (AKM) on Sunday afternoon. The AKM is a large rocket mounted in the after part of the F-3 structure. The rocket is filled with about 600 pounds of solid propellant that, when fired, can circularize the orbit and place the spacecraft into its operational position 22,300 miles above the equator. All of this only works if the spacecraft is precisely pointed in the right direction at the time of the AKM fire. Thus a lot of people were working very hard over the weekend to assure a successful insertion to synchronous. The plan was to fire Sunday evening as the spacecraft was at its 7th Apogee.

The RCA Astro-Electronics Mission Director for the F-1 and F-2 launches had been Jack Frohbeiter. Jack was a highly experienced manager who had started with Astro on TIROS, the first weather satellite. As it became apparent that communications satellites would be a big part of Astro's future, they started to expand the Satcom team and develop new leaders. Jack Frohbeiter was promoted to Program Manager for F-3. The Mission Director for the F-3 launch was Joe Seliga, a telemetry engineer who had worked on TIROS and on the Satcom program from the beginning.   He had a tough job. Besides all of the usual problems associated with a complex communication satellite launch, he also had to deal with the difficulties and complexities caused by the crowded and less than ideal conditions at Vernon Valley.   One of Americom's customer's, Showtime, was desperate to get up on the satellite to compete with HBO. To accommodate them Americom had built a Video Tape Center in the Vernon Valley Earth Station effectively using up all the free space.   As a result the Astro and Americom launch team had their offices and computer facilities in trailers alongside the earth station building.

By Sunday, although the mission had gone reasonably well, there was still some uncertainty about F-3's actual attitude.   F-3's perigee, at 90-miles, was lower than the two previous launches.   There was a question of how much atmospheric drag changed the spacecraft attitude going through perigee.   Jack Frohbeiter was on station in Vernon Valley.   As the time for the AKM fire approached and the final attitude files received there was still some concern about the data.   Three large telemetry monitors were mounted above the Vernon Valley Console. Joe was standing in front of the center monitor with Jack standing behind him.   Engineers sat at the table directly below the monitors making last minute calculations.   The Spacecraft Controller had loaded the FIRE AKM command list into the system -- all it required was a release. As the clock counted down the last seconds before the AKM fire time Joe seemed to be wavering about the decision.   All of us were holding our breath.   A few seconds before fire time Jack shouted "stop..stop -- lock the command list".   All of us let out a big sigh of relief. This too critical an operation to proceed if there was any doubt -- we would have another opportunity tomorrow.

The next day, Monday, was also the first day of the Western Cable TV Conference -- one of the industry's largest gatherings. RCA had a large presence at the show to promote the new Satcom 3 spacecraft that was even now being launched. In fact 11 of the 24 channels on F-3 had already been sold and most of the new customers were at the cable show promoting their soon to come cable offerings. Prominent among these was Ted Turner whose new CNN 24-hour news channel was scheduled to debut on F-3. Some of the RCA personnel manning the Americom booth at the show were a little nervous. If F-3 had launched on time and the AKM had been fired on time, the new satellite would be safely in a geosynchronous orbit by the time the cable show opened.. More nervous then some was John Williamson, Director, Public Affairs, who was RCA's primary contact with customers and the media.

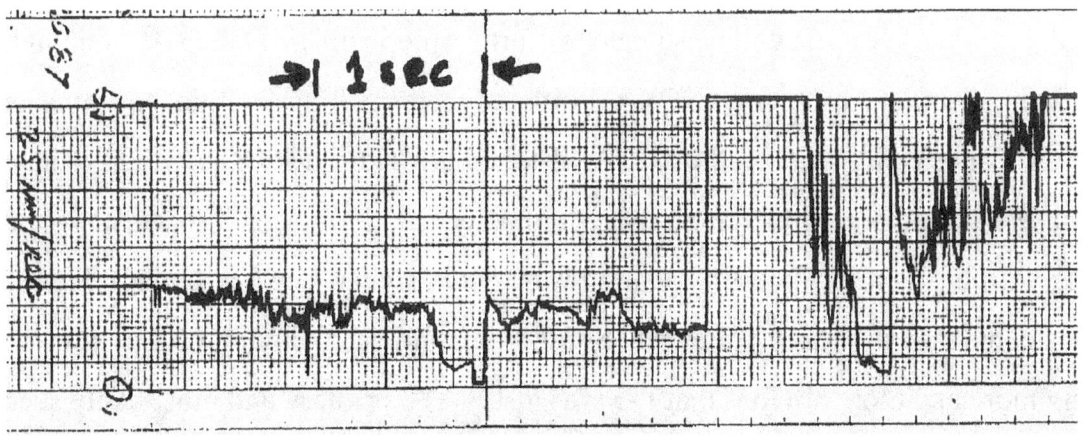

Strip Chart Recording of a temperature sensor near the AKM. The recorder paper speed was 25mm (5 major divisions) per second. The fire command was at "0". About 2-seconds later there is a temperature spike and about 1-1/2 seconds after that the recorder pen pegged and the started scribbling. - *Robert F. Youngblood*

After the AKM firing was scrubbed Sunday afternoon, Astro engineers worked all night gathering more data and refining their AKM plan. By Monday morning the plan looked good to fire the AKM at the 9th Apogee at about 2 PM Monday afternoon. Both Vernon Valley and South Mountain precision tracking antennas were locked on to the F-3 telemetry beacons as the spacecraft approached apogee. Tension was high at Vernon Valley as the critical decision point came near. As F-3 came up on apogee everything looked good and the decision was GO for fire. All eyes were on the strip chart recorder that would display the temperature radiated from the AKM as it fired. The AKM was expected to burn for 26 seconds. The commands were released and the FIRE command executed at 1:52:18 PM EST. The pen on the strip chart started to move down as expected as the temperature increased. At two seconds into the burn the pen showed a jump in temperature and then returned to near normal. At four seconds, the strip chart pen went to max value and then "scribbled" back and forth.

South Mountain reported loss of signal. A few seconds later the Vernon tracking antenna operator reported they had lost lock on F-3. The communication tech who was watching the beacons on the spectrum analyzer reported that both beacons had disappeared. The room was in shocked silence -- no one said a word for a few seconds. Then the tracking antenna operator said "starting box search". The room erupted as everyone started talking about what happened and what to do. A call was made to North American Air Defense Command (NORAD) reporting the problem. NORAD ordered their tracking facilities to start searching for F-3. At Americom's request, the MIT Millstone Hill "Haystack" Radar facility in Massachusets, a high precision system that could detect small objects in space, also started searching. In the next few hours as nothing was found, a deep sense of doom fell on the Astro and Americom team at Vernon Valley and South Mountain. As hours went by it was becoming more and more obvious that F-3 was lost. The primary and most likely scenario was that the AKM instead of burning the total 600 pounds of propellant smoothly for 26 seconds blew up almost instantaneously. What to tell our customers? What to tell the world? There was not much sleep that night for the senior management of Americom. Andy Inglis, Americom President, later recalled "I went home and steadied my nerves with a couple of ounces of medicinal Jack Daniels while waiting for the call that would say: Andy, it ain't so. The satellite has been found." The call never came -- not that night -- or ever!

The next morning, I was in the Vernon Valley Station by 6AM. Many of the Astro and Americom engineers had been there all night. There was no good news -- none of our systems had been able to locate the spacecraft. NORAD's many tracking systems had found nothing. The NASA Deep Space Network had found nothing. Satellite tracking systems all over the world were looking for RCA's lost satellite without success.

At about 7:30AM, John Williamson called me from Las Vegas. "Is it really gone?" he said.

"Well" I said "it certainly hasn't been declared so officially -- but it looks really bad." "What am I going to tell all these guys -- particularly the ones that have bought transponders on F-3?" John asked. "I don't know" I said "but you better get a story together fast. This is no secret -- we've asked just about everybody in the industry for help in finding our lost bird so they all know about it". Shortly after I got off the phone with John, I received a call from an Americom VP telling me to "say nothing to the media -- refer them to John Williamson". That was fine with me. It didn't take the media long to catch on that something had happened -- I started to get calls from newspapers, and TV and radio stations. I referred them to Williamson and they all came back and told me that John was answering "no comment". The New York Times later reported that the TV executives in Las Vegas were "semi-frantic". AP quoted Ted Turner as commenting "Disastrous"

About that time, my secretary reported that there was a TV truck in our parking lot. It was soon joined by a couple others. I had to greet them at our front door and tell them that they couldn't come in and that I had no comment. They kept saying "we know your satellite is lost in space -- you've got to talk to us". Finally, I called my boss in Princeton and told him that we couldn't keep stonewalling -- everybody knew that something happened. Shortly thereafter, I got a call from John Williamson who asked "do you have a copy of the F-1 and F-2 promo film?" When I told him that I did he said that there was an NBC guy out in our parking lot and that I should give him the film because "NBC is an RCA company and we are getting a lot of flack internally to give them the story". When I found the NBC guy, he said "you don't know how big this is -- we are so sick and tired of reporting on Iran and the ayatollahs that we'll jump on almost any other story at this point". That night NBC News repaid our favor with a fairly sarcastic report of the F-3 loss ending with the words "There are a lot of red faces at RCA". As Andy Inglis commented later "As for the red faces, this is simply not true. We were disappointed and dismayed, but we were not ashamed or embarrassed -- nor was there any reason for us to be."

Of course as soon as the other crews found out that NBC had the story there was a great uproar and much righteous indignation and demands to be let into the station. I called John Williamson and told him that they couldn't continue to sit on the story but I also could not have a mob scene in the station while we were still looking for F-3. By this time phones were ringing off the hook throughout Americom from almost every news organization. The Americom switchboard in Princeton was jammed and before the day was over, the public affairs group had fielded more than 200 calls from newspapers, trade publications and broadcasting stations.

John finally called me back in early afternoon and said that RCA had agreed to give a single pool TV crew representing all of the TV networks access to the Vernon Valley station. They would be coming from New York by helicopter and could I please pick them up at the Great Gorge Ski Area heliport. I assured him that I would. At about 3PM I was at the heliport at the rear of the ski area parking lot and could hear a helicopter somewhere in the area although I couldn't see it. I finally spotted the copter at the other end of the valley near the Western Union satellite station flying in circles -- obviously lost. After much arm waving and running back and forth, I got their attention and they landed. The first thing I said to them was "you got pictures of the RCA station didn't you? -- the one just north of the Playboy Hotel -- not the Western Union station at the other end of the valley?" They assured me that they had filmed the right station and of course the lead on all three networks that night showed the Western Union station as the location for "Lost In Space". There was also a shot taken in the Vernon Valley station showing Astro engineer Roger Hogan wiping his face in a gesture of hopelessness -- that clip was rerun on almost every story on F-3 for weeks.

As soon as the F-3 story appeared on the networks, the attention from other media increased drastically. A number of stories appeared blaming "Space Aliens" for the loss of F-3. Almost every day there was a newspaper article updating the story with more speculation on the cause of the loss. Much was made in the media of the fact that F-3 disappeared almost immediately after Ted Turner announced that CNN was going to debut on F-3.

A mournful RCA mascot Nipper, usually listening for "His Masters Voice", listens for a sign of life from RCA's F-3 Satellite in a syndicated newspaper cartoon. - *Don Wright 1979*

On the CBS Evening News with Walter Cronkite on December 11, Charles Osgood read a modified version of "The Night Before Christmas" recounting the F-3 disappearance story.

Cronkite: "The Satcom 3 satellite was launched last week. Its odyssey is the wonder of scientists today and Charles Osgood explains why."

Osgood:
> *"Twas 3 weeks before Christmas and down at the Cape*
> *The Satcom 3 satellite seemed in great shape*
> *It was RCA's baby that NASA would fling*
> *As it happens the 150th thing*
> *To be launched by a delta – a rocket so flyable*
> *It's considered to be absolutely reliable*

It had sent up the Echo the TIROS and more
And Satcom 3's channels – all 24
Would send audio video out of the sky
From a vantage point 22,000 miles high
It cost 20 million before they had loosed (sic) it
Plus 30 more million to insure it to boost it
They launched it last Thursday – they did it at night
Which provided a wondrous and beautiful sight
With its course and direction no fault could be found
As a way or relaying both picture and sound
And it went into orbit and looped around earth
And was tracked by earth stations from Beltsville to Perth
As dry leaves before that before the wild hurricane fly
When they meet with an obstacle mount to the sky
So satellite engines are duly inspired
When their steering or kick motors come to be fired
And so in New Jersey RCA engineers
Hit the button that then theoretically steers
The Satcom to go where they wanted it to
But instead of complying what then did it do
Well the tracking display that the Satcom was on
Showed a very strange thing – that the Satcom was gone
It just vanished it seems out there somewhere in space
And they just couldn't find it around anyplace
On NORAD on COMSAT on NASA and all
From the top of the porch to the top of the wall
Look under the bed or in back of the Moon
And try please to find it and find it real soon
Satcom 3 is still missing – that's the expert's last word
And expensive but now quite extinct sort of Bird
And from somewhere in space comes the seasonal call
Merry Christmas, good night and you can't win them all

Charles Osgood CBS News New York"

Cronkite: "And that's the way it is Tuesday, December 11, 1979. This is Walter Cronkite CBS news – good night."

Almost everyone in Americom was now resigned to the fact that F-3 was gone and that some way had to be found to provide service to our F-3 customers.   The customers were all pretty restless and worried about the future -- many had staked their fortunes on satellite TV distribution. Ted Turner, who had bet everything he had including his billboard business to

22

start CNN, sued RCA to give him two transponders on F-1 -- and he won. RCA agreed to let Turner keep WTBS on F-1 (it was supposed to move to F-3) while RCA appealed the decision. Some F-1 customers who were supposed to move to F-3 now scrambled to reclaim their transponders on F-1. Besides Turner there were other networks that were to have inaugurated service on F-3. Time-Life planned a network featuring BBC programming. There was a new religious channel, a sports a channel and a channel aimed at 50+ viewers. Superstation WOR, New York was subleasing a transponder on F-1 from Showtime and was now concerned about being bumped off if Showtime wanted their channel back. According to AP, Showtime President Jefferey Reiss says Showtime hasn't yet asked for the transponder, but added "We have the right to it, and if we did ask for it, they'd have to relinquish it".

RCA Americom's business was also on the line. Andy Inglis had directed Dennis Elliott, VP Finance, to develop a preliminary financial forecast showing an approximate estimate of the impact of the F-3 loss on the 1980 and 1981 profile. If RCA could not get a replacement for F-3, Western Union and AT&T would steal away their customers and with them RCA's preeminent position as the satellite provider for the cable TV industry. There were empty transponders available on various satellites, however, RCA needed to find all the transponders on one bird so that cable operators would only need one antenna to receive all of their programming.

At Vernon Valley, we regularly used our precision tracking antenna to look at the downlinks of our competition -- Westar and AT&T. I had been reporting to Americom management that the three satellites, that AT&T leased from Comsat General, had very little traffic in them. In fact one of them, Comstar D2, appeared to have no traffic in it at all. Americom management approached AT&T about the possibility of leasing space to put their F-3 traffic on Comstar D2. On February 21, 1980, Americom President, Andy Inglis, announced that RCA had made a deal with AT&T to lease transponders on Comstar D2 for $70,000 per month each and that they in turn would lease these to RCA's customers at the previously agreed $40,000 a month. The potential loss of $5.9 million over the 18 month period of the contract would be covered by insurance on F-3. F-3R the replacement for F-3 was now under construction at Astro and scheduled to launch in November 1981. Americom was back in business!

The F-3 Apogee Kick Motor (AKM) was manufactured by Aero-Jet General Corporation. After the F-3 catastrophic failure, Aero-Jet went of the AKM business. Subsequent AKM solid motors for RCA spacecraft were manufactured by Thiokol. Ultimately in the 1990's, the solid motors were replaced with 100 pound thrust liquid motors built by Royal Ordnance (ARC). They were bi-propellant motors using the same hydrazine fuel as the other smaller thrusters on the spacecraft with nitrogen tetraoxide as the oxidizer for the 100 pounders. With the liquid fueled engines the transition from GTO to geosynchronous orbit was accomplished by a series of small burns over several days -- not one huge one lasting (hopefully) 26-seconds.

Several years later, we received information from the US Air Force that they had found F-3 -- or what remained of it. The USAF has cameras all over the world that photograph the skies looking for foreign objects against the star background. A USAF camera in South Korea had photographed an unknown object and they had made an orbit determination that corresponded to the expected trajectory of F-3 if the AKM had exploded. It was an interesting bit of information but it did not matter much to us anymore -- it was all just history.

Chapter 2 - In The Beginning......

The science author Arthur C. Clarke postulated in 1945 that if a satellite was placed into orbit on the earth's equator and its speed was the same as the angular rotation of the earth, the satellite would appear to be standing still and thus useful for relaying radio signals. The speed required to go into orbit around the earth is about 17,000 miles per hour. The faster the satellite goes the farther it rises above the earth. When a satellite on the equator reaches the speed necessary to match the earths rotation (1 revolution per day) it will be about 22,300 miles above the earth. This is called a geostationary orbit and the one used by most communication satellites. Because the satellite orbits the earth right on the equator (zero latitude) it appears fixed in longitude -- if it is at 78 degrees West Longitude, for instance it would be directly over Quito, Ecuador and due south of Wilmington, North Carolina. Because the satellite appears to stay in the same place all the time, antennas for communication with the satellite can remain fixed in one position -- thus the Dish Network and Direct TV small dish antennas that you see fixed permanently onto roofs in your neighborhood. Communication satellites are now parked in geostationary orbits about every 2-degrees of longitude along the equator around the world. Because of the enormous demand for communications, some of these locations (slots) are occupied by several satellites.

The first truly geostationary satellite was the Department of Defense Syncom-3 built by Hughes Space and Communications and launched on August 19, 1964. It was placed in orbit at 180° east longitude, over the International Date Line. Syncom-3 had two 2-watt transponders one of which was wide band for TV transmission. It was used that same year to relay experimental television coverage of the 1964 Summer Olympics in Tokyo, Japan to the United States. This was the first television transmission sent over the Pacific Ocean.

Syncom-3 had 2-watt transponders and communicated via the omni-directional whip antenna on the top. Because of its low power and low antenna gain, it required large antennas and powerful transmitters on the ground. The angled whip antennas on the bottom of the spacecraft were for telemetry and command.

Intelsat I (Early Bird) was launched on April 6, 1965 and placed in orbit at 28° west longitude. It was the first geostationary satellite for telecommunications over the Atlantic Ocean. It was built by Hughes and was a design similar to the Applications Technology Satellites (ATS) that Hughes was building for NASA at that same time. Tiny by todays standards, it could relay 240 voice circuits or one TV Channel. Designed to operate for 18-months, it actually was in service for four years.

Intelsat 1 - Early Bird - *NASA Photo*

The NASA Goddard Space Flight Center (GSFC) Applications Technology Satellite program experimented with communication satellite technologies starting in 1966. The program consisted of 5 spacecraft built by Hughes and one built by Fairchild. To support this program, GSFC operated three satellite earth stations in Rosman, NC, Mojave, CA and Cooby Creek, Toowoomba, Australia. Although two spacecraft were lost to launch vehicle failures, the program was very successful and yielded an enormous amount of valuable data on the design and operation of satellite communications systems. The experience gained by Hughes in this program provided the foundation for their very succesful commercial communication satellite manufacturing business.

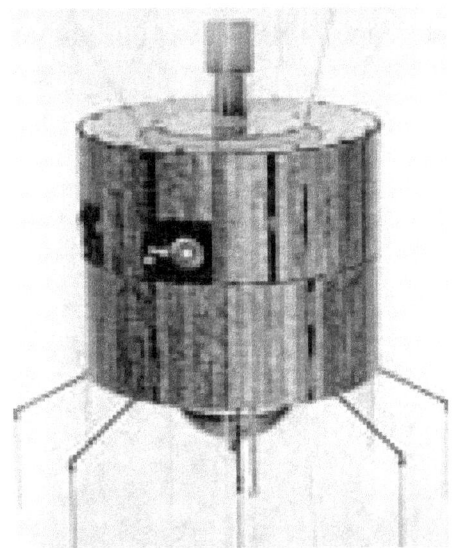

Applications Technology Satellite - *NASA Photo*

Intelsat II F-3, built by Hughes, was launched March 22, 1967, and positioned in synchronous orbit above the Atlantic. The fourth satellite in the series, Intelsat II F-4, was launched September 27, 1967, for commercial operation over the Pacific. The Intelsat II's were twice as large as Early Bird with more than twice the power. They were equipped with an advanced antenna design developed by Hughes that permitted direct contact with a number of ground stations simultaneously. Their design lifetime was three years. Intelsat II F-1 was never operationally useful due to a bad orbit, F-2 lasted for 2 years, and Intelsat F-3 and F-4 lasted for 3.5 years each.

Intelsat III spacecraft, built by TRW, were used to relay commercial global telecommunications including live TV. Three of the 8 satellites in the series (F-1, F-5, F-8) were unusable due to launch vehicle failures, and most of the remainder did not achieve their desired lifetimes. The spacecraft were spin stabilized with a 34-inch tall despun antenna structure. The Main Bearing Assembly (MBA) for the despun antenna would sometimes jam when it got cold causing the antenna to spin up and communication to be lost The solution was to fire thrusters to precess the spacecraft over 180 degrees (upside down) twice a year to keep heat from the sun on the MBA. Of course all communications were lost during this procedure. To my knowledge, after this experience, TRW never built any more geostationary spacecraft.

Intelsat III - Despun horn antenna is at top with the MBA directly below. - NASA *Photo*

While the Intelsat satellites demonstrated the possibilities of international satellite communications, the short life and limited communication capability of the spacecraft made them unsuitable for commercial operations in the North American domestic communications market. A viable satellite communications business required a spacecraft capable of carrying 10 or more TV channels and having a life of more than 7 years. Three companies; Hughes Aircraft Corp., Ford Aerospace and RCA Astro- Electronics were working on designs for commercially viable communication spacecraft in the early 1970s.

The design challenges were substantial. The size and weight of the communication spacecraft was severely constrained by the limited lift capability of the launch vehicles available at that time. The launch vehicle constraints on size limited the size of solar arrays (particularly in the spinner configuration) and thus the power available for the communications payload. Spacecraft thermal design was driven by the spacecraft operating in full sunlight. It also had to be able to survive the twice yearly eclipse seasons when it spent more than an hour each day in the dark. Hughes was the only one of the three companies with experience building and operating communication spacecraft at synchronous altitude -- primarily because of their government supported work on the Syncom and Applications Technology Satellites (ATS) programs. As a result of this experience, Hughes was first company to come to market with a commercial bird, the HS-333.

On November 9, 1972, the first geostationary satellite serving the North American continent, Anik A1, was launched by Telesat Canada. The United States shortly followed suit with the launch of Westar 1 by Western Union on April 13, 1974. Both of these satellites were Hughes HS-333 spin-stabilized spacecraft that were based on their earlier designs for NASA and Intelsat. The major improvements were a despun shaped beam antenna that concentrated the communication signals on the desired geographic area and a 7-year design life. The body of the spacecraft spun at 100-rpm and an electric motor despun the parabolic reflector antenna to keep it pointed at the earth. The pointing reference for the antenna was a pilot radio signal transmitted from the ground -- equipment on the spacecraft kept the antenna pointed at the pilot signal while the body of the spacecraft spun at 100-rpm. If something happened to the pilot signal, the spacecraft antenna would spin up and communications would be lost. Because this problem had occurred a number of times with the Canadian Anik birds, Hughes later modified the HS-333 to reduce reliance on the pilot signal. The Westar spacecraft were equipped with a Backup Despin System (BUDS) that used an earth sensor on the spacecraft to assist in maintaining despun operation if the pilot was lost.   Although the HS-333 was nominally a 12 channel spacecraft, 2 channels had to be turned off during the two, 40-day eclipse seasons because of power and thermal constraints.

Westar-1, an HS-333 spacecraft, atop a Delta 2914 launch vehicle being prepared for launch. The eliptical mesh antenna (despun section) is atop the cylindrical body that is covered with solar cells.   One half of the protective shroud is in place - *NASA Photo*

Anik and Westar each operated three Hughes 333 spacecraft and all of them exceeded the design lifetime. Both companies experienced occasional communications outages because of antenna spinups. In one instance, a lightning strike at the Allan Park, Ontario facility that transmitted pilots for all of the Anik spacecraft caused the antennas on all three spacecraft to be spun up simultaneously. My conversations with the Telesat Canada personnel about that incident proved once again that they were very good at what they did and recovered all three spacecraft relatively quickly. However, they all felt that life would have been a lot easier had their spacecraft been equipped with BUDS. I did understand that after that triple spinup incident they relocated their pilot facilities to three separate geographic locations.

In December 1975, RCA Glōbcom was using two transponders on Westar 2 to carry their Alaskan communications traffic until RCA's spacecraft could be launched. During the RCA Satcom 1 launch operations, Westar 2's antenna spun up and RCA Glōbcom lost communications between New York, San Francisco and Alaska. Based on the performance of the Canadians when the Aniks spun up, we expected that the RCA's communications would be restored in a few hours. It took Western Union 2-1/2 days to to get the antenna despun and locked back on the earth. Jim Judson, the Westar Operations Manager, (and a former colleague of mine at RCA) called me up in the middle of the event complaining bitterly that their troubles had received widespread press attention and that I was the leaker. I told him "Jim, when you dump RCA's traffic at a time when all the media is focussed on RCA's launch, you might expect that you'd get caught". Of course, I had probably contributed to Jim's agida by hiring away some of Westar's best employees.

The RCA Astro-Electronics Division in Hightstown, NJ was one of the pioneers in space. They had designed and built the highly successful TIROS and DMSP weather satellites and the Ranger Lunar Probes and the RELAY low orbit communication satellites for NASA. The project to build the Satcom spacecraft for RCA Glōbcom, however, was their first geostationary communication satellite project. It was also Astro's first commercial project (all previous work had been for the US government) so it had top priority for talent within the company. The Glōbcom performance specifications for Satcom required a design on the leading edge of communication spacecraft technology at that time. The design was driven by the fact that there are fixed costs in a satellite project independent of the size of the spacecraft. If those costs are spread across a 24 channel spacecraft instead of a 12 channel spacecraft the individual

31

channels can be sold to customers for much less. Glōbcom knew that a cost-effective spacecraft for the competitive North American market would require a 24-channel spacecraft that could operate through all seasons and be able to provide services for at least 8 years. Two of the Astro designers of the satellite, Jack Kiegler and C.R. Hume, published a paper "Maximum Communication Capacity for Unit Cost" describing the Satcom spacecraft and its economic advantages. It was truly an innovative design.

The only problem was that such a spacecraft would probably weigh about 2000 pounds and the existing Delta 2914 launch vehicle could only deliver 1593 pounds to Geocentric Transfer Orbit (GTO). RCA looked at the possibility of using the Atlas Centaur launch vehicle, that could put 4200 pounds in GTO, to launch two satellites at a time. The problem with that was if there was a launch vehicle failure, RCA would lose two spacecraft (and our insurers would have fallen on their swords). At that point, the single remaining ground spare spacecraft would have to be launched on another Atlas which would be a very expensive one-satellite launch. RCA then embarked on a project to pay McDonnell-Douglas to redesign the NASA Delta 2914 Launch Vehicle so that it could lift the heavier load of the new Satcom spacecraft. This was the first time that a commercial company had redesigned a NASA launch vehicle to increase its capability. The solution was to replace the Thiokol Castor II solid propellant strap-on motors with much larger Castor-IV's. This was not a trivial modification. The strap-on motors had to be dropped off the launch vehicle as they burned out to reduce weight. The Castor IV's were 13-feet longer than the II's and required an upper support point on a strengthened Liquid Oxygen (LOX) tank. The antislosh baffles in the LOX tank had to be redesigned for the changed vibration frequencies. A new separation system to push the released Castor's safely away from the Delta was required. The now heavier launch vehicle required larger hoists on the launch pad and revised down-range procedures. The redesigned vehicle was designated the Delta 3914 and was subsequently used for other commercial satellite projects with the users paying fees to RCA.

All of that work really paid off. When the Satcom's became operational in early 1976, our competition was Western Union flying 12 channel Hughes 333's and AT&T flying 24 channel Hughes 351's. Although AT&T had a 24 channel bird, it required launching on an Atlas Centaur -- a very expensive launch! RCA could charge $40,000 an month for a channel where AT&T was charging $60,000. RCA ended up with more available channels than Western Union and lower cost channels than AT&T.

RCA's entrance into the commercial communication satellite business was driven by requirements to improve communications in the state of Alaska. In 1971, RCA bought the Alaskan Communications System from the federal government for $125-million. The system consisted of community telephone systems interconnected by microwave, VHF radio links, submarine cable and the US Air Force-built White Alice tropospheric scatter system. The system was woefully inadequate to serve the rapidly growing Alaskan economy. Much of Alaska, particularly the bush communities had little or no communication to the outside world. An Alaska pioneer said this about TV: "Live television, a given anywhere else in the U.S. arrived late in Alaska. Entertainment programs were a week or two late arriving in Anchorage by film or tape. After showing in Anchorage, the material was sent onward for even later showing in Fairbanks and then Juneau. National news was taped off the air in Seattle and put on the first available northbound plane. In most cases, Walter Cronkite addressed his Alaskan audience a day later than the Lower 48." RCA realized that the only way to provide a telephone, TV and data communications for Alaska was with a satellite system. In 1972, RCA filed an application with the Federal Communications Commission (FCC) to construct and launch a four satellite system at a cost of $256 million to serve Alaska, Hawaii and the Continental United States.

White Alice tropospheric scatter relay station on an Alaskan mountaintop. The diversity system required two large antennas on each path. This complex network was replaced by satellite earth stations in each of the 20 areas served by White Alice - *US Air Force Photo*

RCA did not wait for their own satellites to be built to start to improve Alaska communications. On April 5, 1973 RCA Global Communications Inc. (Glōbcom) signed an agreement with Telesat Canada to lease two transponders in their Anik 2 spacecraft. RCA used these two channels to provide telephone and television traffic between New York City, San Francisco and Alaska thus becoming the first American domestic satellite provider. The system went into partial service on December 20, 1973.

On January 9, 1974, RCA Glōbcom gave a demonstration to the press of telephone communication between New York, Washington DC, San Francisco and Anchorage, Juneau and Nome, Alaska. RCA Chairman Robert W Sarnoff made the first call and his voice gradually faded away and then slowly returned. Although embarrassing, it turned out to be a telephone problem that was quickly corrected. RCA Glōbcom President J. C. Hepburn then called California Lt. Governor Ed Reinke.   Senator Mike Gravel in Washington called Alaskan Governor William Egan in Juneau.   Glōbcom Executive VP Howard Hawkins announced that the satellite system would make communications costs 25% lower.   He said that a New York to San Francisco voice circuit that now costs $2566 would cost $1700 in the satellite system.

After Western Union launched their satellites in 1974, they complained to the Federal Communications Commission (FCC) that it was unAmerican for a US company to use a Canadian satellite.   Although Western Union was not licensed to serve Alaska, they argued that any company providing Domestic US service ought to do so using an American bird. Western Union prevailed in court and on May 2, 1975 the RCA Glōbcom traffic was transferred from Anik-2 to Westar 2.   At that time I was working for American Satellite whose traffic was also on Westar 2.   I participated in the traffic transition conference at Western Union when the RCA switchover was accomplished.   I got to know some of the Western Union people who I would later recruit to work for RCA.

Bartlett Earth Station near Talkeetna, Alaska with Denali (Mt. McKinley), the highest mountain (20,335 feet) in North America, in the background.   The Bartlett Earth Station, built in 1969, was a former Intelsat "A" station and was approximately equidistant from Fairbanks and Anchorage, Alaska and was linked to both. It was a typical Intelsat A station with a 30 Meter, 15-Meter and 5-Meter dish antennas.   Until RCA provided satellite service to all of Alaska in 1974, this station provided the only direct satellite link between Alaska and "the outside world". - *ptt.ak.com Photo*

Chapter 3 - The Launch Is When?

RCA Glōbcom's fixed price contract with RCA Astro-Electronics Division required Astro to deliver a spacecraft in 24-months to be launched in December 1975 from Cape Canaveral, FL. Astro also had to deliver two Telemetry, Tracking and Command (TT&C) stations four months before the launch. Each of these identical TT&C stations contained the hardware and software for launch and in-orbit control of three spacecraft. Globcom for their part was committed to providing two communication earth stations with 13-meter dish tracking antennas; one on the east coast and one on the west coast to house the Astro-supplied TT&C systems. All of this had to be accomplished in time to support a December 1975 launch This was a very ambitious schedule but was absolutely necessary if RCA was going to be a real competitor in domestic US satellite communications. RCA and all of their subcontractors had their work cut out for them.

RCA Glōbcom had a great deal of experience in satellite communications. They had built satellite earth stations and microwave radio links in many parts of the world. In 1972, Glōbcom received a contract to build an Intelsat Standard A Satellite Earth Station with 30-meter dish antenna and microwave facilities in Beijing in advance of President Nixon's historic visit to China. They built and operated an Intelsat-A station on Guam. They had also built the earth stations and microwave facilities to support the Alaska communications traffic going to San Francisco and New York.

Now they were faced with developing earth station facilities for all of the cities to be served by their new satellites. Of primary concern of course were the two stations to contain the east and west coast Tracking, Telemetry and Command (TT&C) facilities. Glōbcom designed an earth station building for the TT&C systems that also included space for the communication facilities required to serve a major city. These facilities included high power transmitters, receiving systems, telephone multiplex equipment and television transmission and monitoring equipment and emergency power systems. Because of possible radio interference problems, satellite earth stations were usually located in rural valleys shielded by surrounding mountains and connected to the nearby city by microwave radio links. The west coast TT&C was to be in an earth station serving Los Angeles. Land had been obtained for an earth station site in Coyote Canyon near South Mountain in Somis, Ventura County California. Approval had been obtained for a 3-hop microwave link between the South Mountain Earth Station and the RCA Central Telephone Office on Wilshire Boulevard in downtown LA. Construction of the earth station had been started earlier in the year and was on schedule. As with many land deals in the west, however, RCA did not own the water rights for the property so could not drill a well. Two 110,000 gallon vinyl water bladders were installed in pits on the hill above the station to hold fire and flushing water. Every few weeks a tanker truck from the Las Posas Water Company replenished the water supply on the hill and brought in bottled potable water.

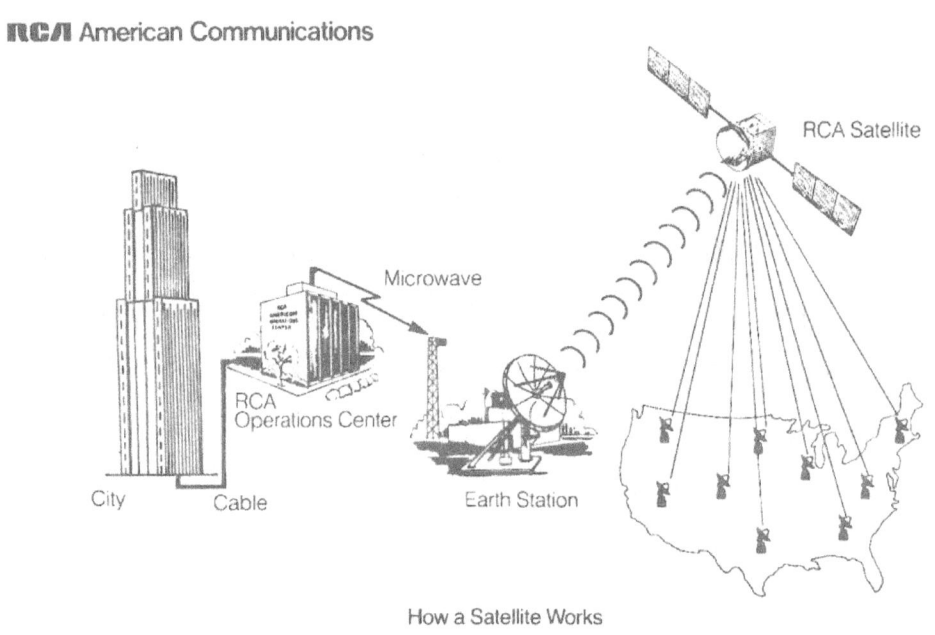

A basic satellite system showing customers in the city connected to the RCA Center by telephone local loops or video cable. The circuits are then carried by microwave to the earth station, which is in a rural area, and transmitted up to the satellite. The satellite broadcast covers the whole US. - *RCA Americom*

Glōbcom was having some problems deciding where to put the east coast facility. Part of the problem was that a satellite earth station can't be placed just anywhere. Satellite communications use the same C-Band frequencies (6 GHz uplink and 4 GHz downlink) used by terrestrial microwave system. To prevent interference between the satellite and terrestrial systems, earth stations have to be located in electronically quiet areas. This usually means valley locations away from major urban areas. Once a location has been chosen, an FCC required field survey and frequency coordination study must be completed to rule out interference. Because the earth station is in a rural area and the customers are in cities, the two locations must be interconnected with terrestrial microwave (11 GHz) facilities. The objective then, in locating earth stations, is to find the closest electronically quiet area to each city to minimize microwave interconnect costs.

The RCA Glōbcom Central Office in 60 Broad Street New York was connected by a 4-hop microwave link to a small earth station facility located in King of Prussia, PA (Valley Forge) near Philadelphia. The station was located on land leased from GE adjacent to a GE plant. The existing RCA building wasn't suitable for a TT&C facility nor was there sufficient land available to construct a suitable building. Because of these problems, the initial plan was to put the east coast TT&C in an earth station to be built in southern Maryland to serve Washington, DC. It was thought that there would be more than enough Private Leased Channel (PLC) telephone business from the Federal Government in Washington to support a large earth station. Land had been obtained in the rural town of Benedict near the Patuxent River in Charles County and preliminary land use permits had been obtained. Frequency coordination had been completed for the whole 500 MHz C-Band satellite communications spectrum. At the same time, Glōbcom marketing people were surveying the PLC possibilities in Washington. What they found was, that despite deregulation, AT&T had such a solid lock on US Government business that breaking in to it would be virtually impossible. It was now early 1975 and the F-1 launch was scheduled for December 6, 1975.

If Washington wasn't feasible, the best alternative would be to locate the new earth station so that it could serve New York. The existing small Valley Forge station could then be used to serve the Philadelphia/Camden area. The logical location to serve New York was Vernon, NJ in Sussex County where Western Union and American Satellite already had earth stations. Vernon was in a valley surrounded by the 1400-foot Waywayanda Mountains providing a radio-quiet site. The Vernon Township Committee thought that earth stations were great ratables and made it as easy as possible for RCA to locate there. The town told RCA Glōbcom that they would treat them as a utility and that ,under their land use ordinance, utilities were a conditional use in all zones. That ruling meant that RCA could build the earth station almost anywhere in the Vernon Valley. While that ruling seemed great at the time, and allowed RCA to move quickly, it would ultimately cause the company major problems with its earth station neighbors. RCA bought 10 acres from local dairy farmer Sam Edsall and applied for use permits which were quickly granted. Ground was broken in late May and construction started at a frantic pace. Glōbcom had contracted a highly experienced construction company and given them incentives to make the building habitable enough so that the Astro TT&C equipment could be moved in by August 4. The contractor worked three shifts and through weekends to make the schedule. At one point the Carpenters Union walked off the job just as the forms for the concrete pedestals for the steel work had to built. The project manager determined that clay flue tiles of the correct size could be used for forms without the carpenters help and the job went on.

By August 1, the shell of the building was completed, it was weathertight and utilities including electric were available. Most internal walls were not complete but the raised computer floor had been installed in the operations room area. Vernon Township issued a conditional Certificate of Occupancy so that we could move in. Astro delivered the TT&C Control Console and equipment racks and the contractor built a clear plastic enclosure around the operations area. Room air conditioners were installed through the plastic for environmental control. The contractor continued to construct the building around us as we worked. It rained heavily that summer and the area around the station became a sea of mud. A couple of laborers were assigned full time to clean up the tracked in dirt and to shake out the door mats that had been placed all around. We also changed or cleaned equipment air filters every day. Within a week or so, the TT&C system was up and running and system checkout started -- or so I thought.

As we soon found out, although the hardware had been delivered on schedule, the system software was still a long way from being ready.   The analysts and programmers of the Astro TT&C Software Group arrived with the system and were going to be with us for quite a while until the software was ready to support a launch.   I was chagrined to say the least because I had planned on using the system for familiarization and training of the Spacecraft Controllers and Analysts. Instead it would be tied up for two shifts each day for software development with tests sometimes running on the third shift.

Each TT&C Facility required a Precision Tracking Antenna to gather accurate range and azimuth and elevation pointing data in order to determine the spacecraft orbit.   The antenna also had to be able to move fast enough to track the spacecraft when it was in transfer orbit immediately after launch. Glōbcom had bought 13-meter diameter dish antennas mounted on 25-foot pedestals from the Datron Corporation in California.   The antenna had been delivered to Vernon Valley in a large number of big boxes.   The plan was that Datron would bring in a team of their installers from California to erect the antenna.   We immediately ran into a union problem.   The construction of the earth station building was completely union organized. The union trades blocked the Datron team and the Iron Workers Union declared the job was theirs.   The first day that the Iron Workers were on the job I arrived at about 7:30AM to find the foreman sitting on the antenna pad having his first beer of the day -- not an encouraging development.   The iron workers dogged the job. Everything seemed to take at least twice as long as it should.   One day when we had an expensive rental crane on site, we weren't getting anything done because the iron workers were on a slowdown.   The RCA Project Engineer, Walter Braun, blew up and told the iron workers that if they didn't start working they were going to pay for the crane overtime.   They grudgingly got to work. The other union trades on the job were getting really pissed because they thought that the iron workers were giving all the union guys a bad name. They elected a 6-foot, 5-inch brawny electrician nicknamed "Moose" to read the iron workers the riot act. Moose told them that if they didn't shape up, the other union guys were going to kick their asses. Miraculously the iron workers became productive and the antenna got erected on time. Fortunately, the Datron Tracking Antenna was completed at South Mountain by Datron's installation team without incident.

Slowly but surely the Vernon Valley building was finished, grounds were landscaped, the parking lot was paved and we were able to achieve some semblance of a normal operation. The software people were still with us and had been joined by some of the Astro Launch Team. There were also technicians for communications, power and HVAC contractors on site clearing "punch list" items or conducting tests. There were still many deadlines to be met and obstacles to be overcome but at least the two TT&C facilities were pretty much complete and RCA Glōbcom Spacecraft Operations had a home.

Vernon Valley TT&C June 1976 - John Bailey and Guy Crowther, Spacecraft Controllers at the TT&C Console, the author, standing, is on the network phone. Computer Tech "TP" Tubbs is at the TV monitor and Computer Analyst Irv Harrison is at the equipment racks. Note the large telemetry displays above the console. - *RCA Americom Photo*

Chapter 4 - AS-1000 Spacecraft

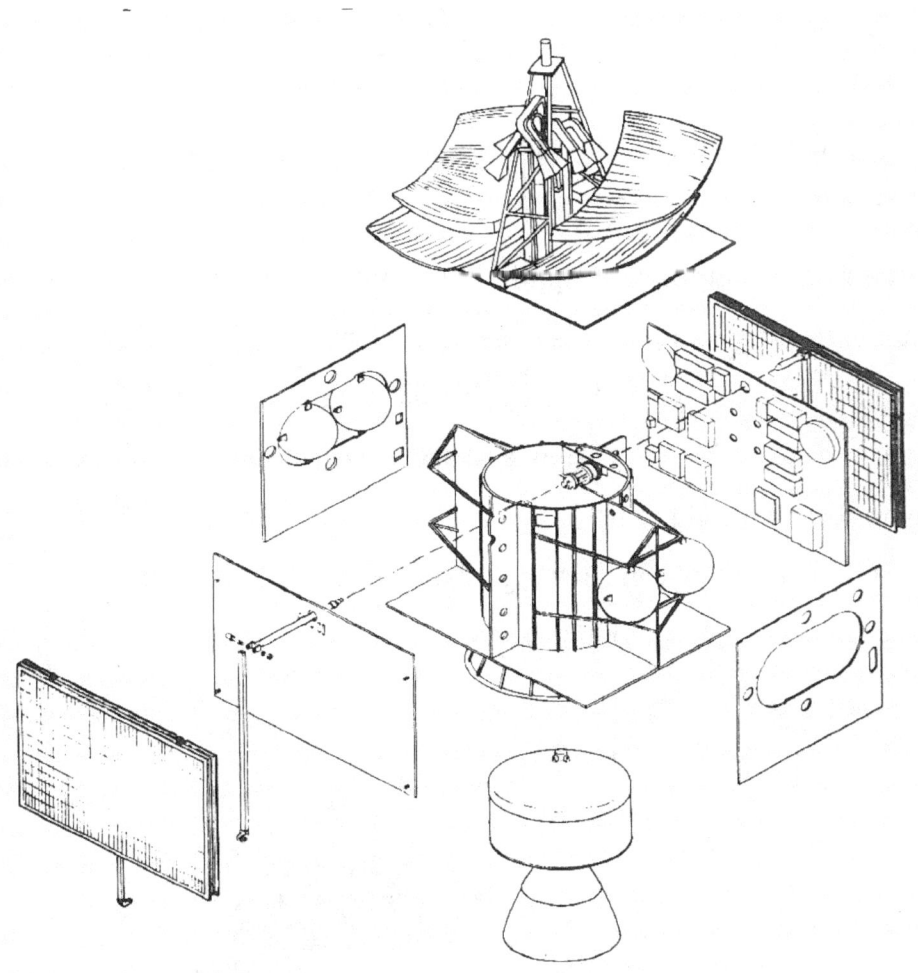

Exploded view of the AS-1000 spacecraft.   The central column and cruciform assembly carry the launch loads to the launch vehicle interface.   The apogee motor (bottom) is mounted within the column which extends up to and provides a rigid base for the earth facing antenna. The central structure also supports the four spherical fuel tanks.   Most of the electronic components are mounted on the North and South panels.   The solar arrays (bottom left) are folded down on these panels. - *RCA Astro-Electronics Division Illustration*

In response to RCA Glōbcom's specification, RCA Astro- Electronics Division proposed to build a new Communications Spacecraft, the AS-1000. Designing and building a communication spacecraft, particularly a completely new type, is not a trivial matter. A communication satellite is a complex electronic system that must live in a hostile environment in space for a long time.   The biggest design constraint is that the maximum total weight of the satellite is controlled by the "throw weight" capability of the launch vehicle. In the case of the AS-1000 spacecraft, that was to be launched on a Delta 3914, the total weight including the spacecraft itself, the solid rocket AKM motor and fuel for eight years was 2000 pounds. Unlike the earlier 12 transponder "spinner" spacecraft built by Hughes, the AS-1000 would basically be a box stabilized on all three axis so that its antenna would always point at the earth.

Astro's goal was a weight optimized design for maximum communication capacity. To meet that goal meant that the spacecraft structure had to be constructed with lightweight metals like beryllium, titanium and aluminum and composite materials such as kevlar and carbon-carbon. Structural parts are made as thin as possible and had holes cut in them and excess metal machined away to reduce weight as much as possible while maintaining the required strength. Every component, box and device had a weight budget and the total budget for all of the component parts could not exceed 2000 pounds.

The AS-1000 box structure standing on the round launch vehicle interface. The panel facing left is the North panel. The four holes in the center of the panel are for thrusters 1, 2, 3 & 4 that are used for North/South maneuvers. The round hole at the top of the panel is for the solar array shaft. The panel to the right is the East panel. The large hole is for two of the four spherical fuel tanks. The four round holes are for thrusters 5, 6, 7 & 8. The two small square holes are for the Sun and Horizon Sensors. - *RCA Astro-Electronics Photo*

The spacecraft has to operate in a complex thermal environment.   The temperature in outer space is about -270 degrees Celsius (C).   At the same time, for most of the year, the spacecraft receives warming energy from the sun.   Because the spacecraft orbits around the earth, the sun shines on different sides of the spacecraft throughout the day.   At local morning it shines on the spacecraft east face, at noon on the base plate, at dusk on the west face and at midnight on the earth (antenna) face.   The energy from the sun also changes seasonally. During northern hemisphere summer, the sun shines on the spacecraft north face -- in the winter it shines on the south face. In the vernal and autumnal equinox seasons the spacecraft goes in the shadow of the earth (eclipse) for up to 72 minutes centered around spacecraft midnight and the spacecraft is exposed to an extremely cold environment and must run on battery power.

Even with these varying external temperatures, the spacecraft internal environment must be designed to accommodate the limited allowable temperature ranges of the components. The hydrazine fuel used in communication spacecraft freezes at about -2 degrees C.   The Nickel-Cadmium batteries used in the spacecraft should operate no colder than -10 C.   Individual electronic components cannot be operated at extremely cold temperatures. Conversely, there are electrical components, and rocket thrusters in the spacecraft that generate heat. For instance, the Traveling Wave Tube Amplifiers in each of the communication transponders require about 15 watts of electric power and transmit a 5-watt radio signal -- the remaining 10 watts is dissipated in the spacecraft as heat . The thermal design involves trying to keep all of the components in the spacecraft in their thermal "comfort zone".   Components that are excessively hot are mounted so that their heat can be conducted where needed or radiated into space. Components that are cold can be located to use heat from other components or the sun. For some components such as batteries and fuel system tanks and plumbing, electric heaters are required. It is a complicated balancing act.

All of the electrical power for the spacecraft comes from the sun.   As the spacecraft orbits the earth the solar array is driven in the opposite direction by a one-rotation/day clock motor so that the array always faces the sun. The solar array provides sufficient power to operate all of the on board equipment and also keep the batteries charged. During eclipse season, or if the solar array is turned away from the sun for some reason, the batteries provide all of the power for the spacecraft. The batteries provide some unique thermal challenges. When the batteries are discharging they are exothermic and get very hot -- this heat must be radiated into space. When they are being recharged they are endothermic and get cold enough to require electric heaters to maintain their ideal operating temperature.

When in orbit there are forces working on the spacecraft to change its orbit and cause it to move from its assigned station. These include the gravitational acceleration of the sun and moon, solar pressure and the effect of the earth not being spherical. Corrections to the spacecraft orbit are made by firing thrusters in maneuvers to correct latitude and longitudinal errors. About 85% of the on-board fuel is used for North/South (latitude) maneuvers and the remainder is used for East/West (longitude) maneuvers and a small amount for attitude corrections. The AS-1000 had twelve 0.1-pound thrusters mounted on the East, West and North faces of the spacecraft.

In order to provide optimum communication performance the spacecraft antennas must stay pointed directly at the earth at all times. An earth sensor scans two chords across the circular earth to provide a reference for the attitude control system. A momentum wheel spinning at 6000 RPM provides control of spacecraft pitch. Magnetic torquing coils working against the earth's magnetic field provide control of roll and yaw. Should roll/yaw exceed the capability of magnetic correction, thrusters are used. Thrusters are also used to unload the momentum wheel so that it operates in its optimum speed range (about 6000 RPM)

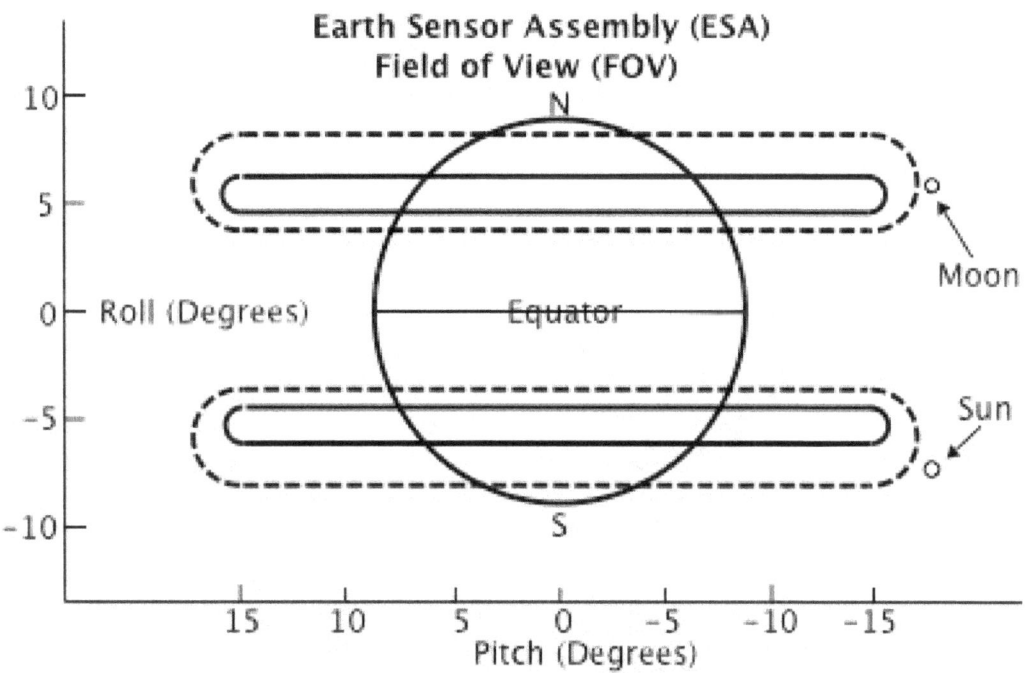

Earth Sensor Assembly (ESA) Field of View (FOV)

The Earth Sensor Assembly (ESA) scans the earth to provide a reference to keep the spacecraft pointed towards the earth. The solid line scan shows the advertised Field Of View (FOV) for the ESA. The dotted line shows the actual FOV that we determined once F1 was on orbit. The diagram also shows how light from the moon and sun could encroach in the ESA because of the larger than expected FOV. - *Robert F. Youngblood*

During transfer orbit, the spacecraft (with the solar arrays folded against each side) spins at 60 RPM. This requires that all of the spacecraft weight must be balanced equally about the spin axis. If not, lead weights have to be added to spin-balance the spacecraft (much like balancing a tire). Of course, every pound of lead added means that one less pound of fuel can be carried. Great care is taken to try to place all of the spacecraft components so that they are balanced around the center axis of the spacecraft and that only a minimum amount of lead is required.

Cut away view of an AS-1000 Spacecraft with solar arrays folded inside the protective shroud of the Delta 3914 Launch Vehicle. This view also shows the overlapping "petals" of the communications antenna reflector and the radio frequency feed horns on the tower at the center of the antenna. When the Delta is above the earth's atmosphere, the shroud splits and falls away to allow the spacecraft to be separated from the launch vehicle. - *RCA Astro-Electronics Division Photo*

The construction of spacecraft is extremely detail oriented. Every single piece part that goes into the spacecraft has to be "flight rated". That means either that the component has flown successfully in another program, was manufactured to military space qualified specifications or exhaustively tested by Astro in a space environment. Then as qualified components are assembled into "black boxes" and the boxes into subsystems they are thoroughly tested. They must be able to operate in the vacuum of space over the temperature ranges expected and survive the severe vibration, shock and acoustic loading experienced during launch. Astro had extensive testing facilities including thermal-vacuum chambers of many sizes and various shock, acoustic and vibration facilities.

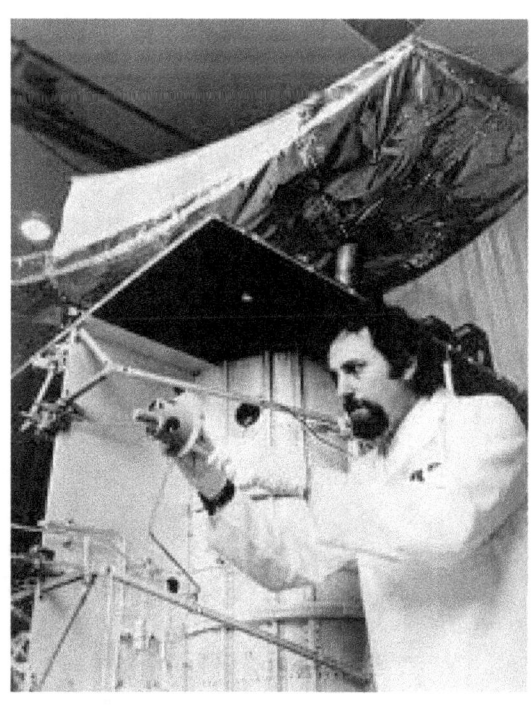

F-1 under construction. The technician is installing one of the thrusters used for East/West maneuvers. Note the fuel line going from the thruster into the central column of the spacecraft. Note also the vertical and horizontal petals of the communications antenna at the top of the picture. - *RCA Astro-Electronics Division Photo*

Components that fail testing are rejected for flight until they are reworked, retested and pass. Because of the schedule pressures to complete F-1 in time for a December launch, Astro was using F-2 (scheduled to launch in March) as a "hanger queen". That is, if a card or box failed in F-1 testing, the same component would be removed from F-2 and installed in F-1. The F-1 component would be reworked and retested and then installed in F-2. F-2 was subsequently launched with the F-1 reworked components. Some of these reworked components subsequently failed in flight causing us untold grief and sleepless nights.

The three AS-1000 spacecraft were identical in all respects. No matter where they were located in longitude along the North American arc of the equator, the antenna beam could be pointed to cover the 50 states by inserting an offset angle in the pitch control loop. For instance, when F-1 was located at 135 degrees West Longitude, it had a 2.45 degree eastward pitch offset to optimize the beam for the lower 48, Alaska and Hawaii.

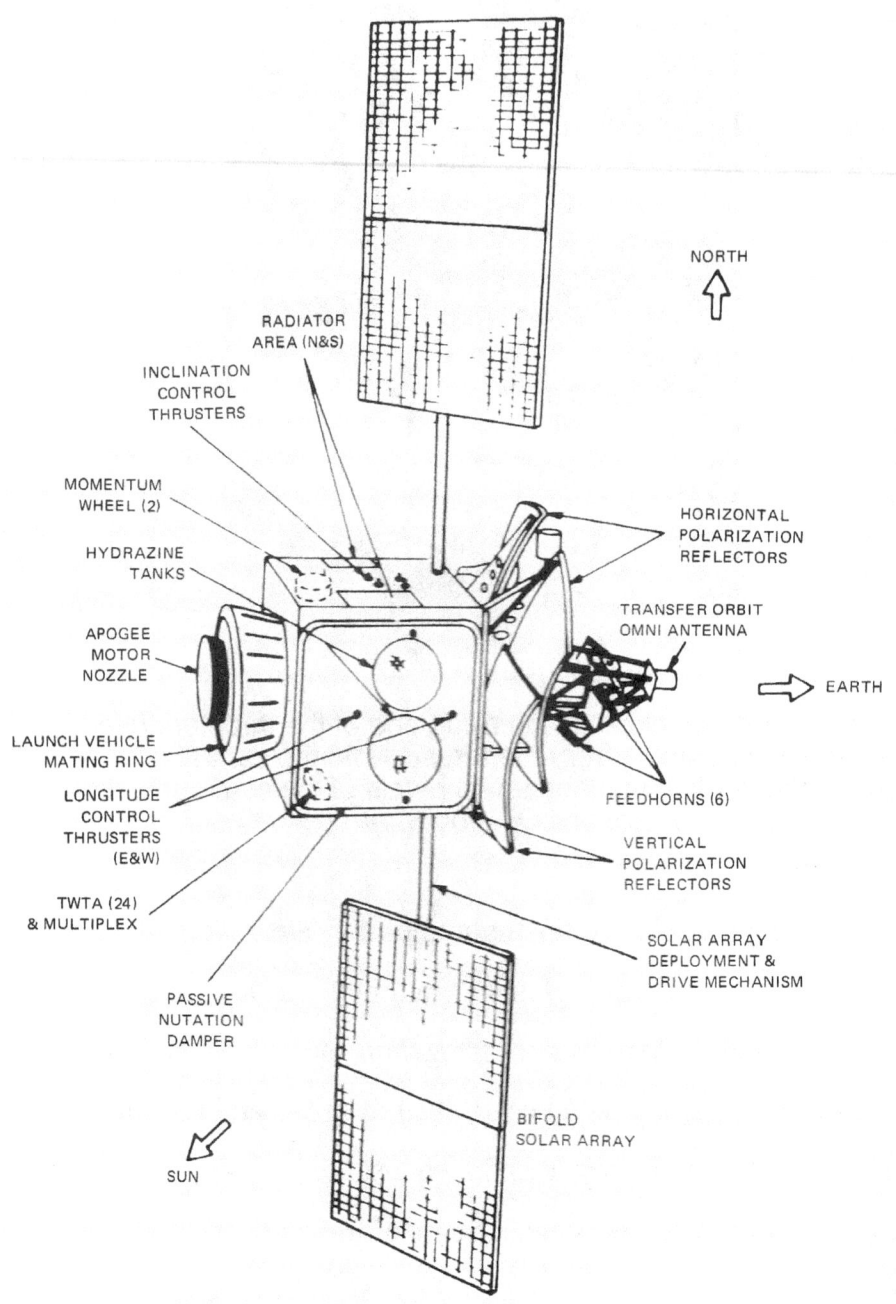

RCA Astro-Electronic Division AS-1000 Spacecraft for SATCOM 1, 2 and 3 - *RCA Astro-Electronics Division Illustration*

Chapter 5 - Team Building

In June of 1975 when I came back to work for RCA, virtually nothing had been done to set up a Spacecraft Operations organization and the launch was only six months away. Ray Balon, who I knew as the Nimbus Control Center Manager at Goddard, had been on board for about a year and was acting as Spacecraft Operations liasion to Glōbcom Spacecraft Engineering. RCA Glōbcom's main office was at 60 Broad St near Wall Street in New York City. The Glōbcom Spacecraft Engineering group was located in temporary offices in Princeton Junction, NJ to be closer to RCA Astro at Locust Corners in East Windsor, NJ where the spacecraft was being built. Ray was scheduled to be my manager of the Telemetry, Tracking and Command (TT&C) facility at South Mountain, CA. As soon as I was on board at Princeton Junction, he was going to California to start recruiting staff.

The day in May that I got my offer letter to return to RCA, I was working at the American Satellite earth station in Vernon, NJ. I got in my car and drove over to the new RCA site on Edsall Road. It was pretty discouraging. Grading was complete but not much else. The site was on the side of a hill and the contractor had cut into it and pushed the dirt down hill to make a level area for the earth station. Concrete footings for the building had been poured and I paced them off trying to compare them with the drawings I had seen during my interview in New York. There was a construction trailer near the road but no sign of activity. All I could think was "we'll never make it!"

Glōbcom Spacecraft Engineering was located in an old commercial building next to the Princeton Junction Amtrack station. John Christopher, who Glōbcom had hired from the General Electric Space Division, was Director, Spacecraft Engineering. John had designed the GE Japanese Broadcast Satellite (JBS) and suggested to Astro that they adopt a similar configuration for Glōbcom's new bird. He had built up a staff from recently laid-off Astro engineers and new hires. John and his staff were working closely with and monitoring Astro to be sure that the spacecraft would be completed on time and in accordance with Glōbcom specifications.

Princeton Junction was only about a 10-minute drive from the RCA Astro-Electronics plant in East Windsor. When I arrived at the Princeton Junction office, Ray Balon briefed me on the status of the satellite control equipment that was to be installed in Vernon Valley and South Mountain. Ray said that the hardware seemed to be working well but that the software development was far behind. We went over to Astro to see how things were going.

Astro was in a large building with office spaces, labs, environmental test areas, machine shops and production areas. As we entered the building, large sections of the building were dark. We walked past empty offices and lab spaces until we came to the TT&C area -- the only lighted area in the whole wing. The space business had not developed quite as fast as RCA had anticipated -- the reason for the recent layoffs. The new communication satellite business would soon change all that. In a few years, after Astro built their business on our success operating their satellites, all of these spaces would be filled to overflowing.

The hardware for the Vernon and South Mountain TT&C systems were set up in a large room just as they would be installed in the stations. We were given a briefing by Astro management that was optimistic and upbeat, however, it was obvious to me that the software was a long way from where it should be at this point in the program. Admittedly, this was a challenging program. The computer system consisted of two Hewlett Packard 2100 mini-computers. The Data computer processed the telemetry data from up to 3 Spacecraft. The Control computer contained all of the commanding functions and the operator interfaces and display functions. Today all of those functions would not be hard at all for a single laptop PC but this was 1975 -- the HP-2100 had only 32K of memory! That meant that in a real time system, such as ours, program overlays or pages had to be swapped in and out of memory every time a program was used.   Only one small program called Master Cycle containing the system clock and the overlay control was resident in memory at all times.   And the disk drive where all the programs and everything else was stored had only a capacity of 20 Megabytes!   The primary reason for selecting these HP computers was that they operated under the HP Real Time Executive (RTE) operating system.   The HP-2100 was the only proven real time mini-computer system available.   HP had designed it for such mundane uses as monitoring manufacturing production lines -- we were going to use it to handle real time data simultaneously from 3 satellites.   I used to kid about the fact that if a real time system didn't work right all the ones and oughts ended up on the computer room floor.

I was quite concerned about the state of the software.   It was not only that the program was behind schedule but also by the fact that we were behind schedule on a project that was seriously testing the capability of the available computer resources.   There appeared to be many opportunities for things to go seriously wrong -- and we were only six months away from launch.

I also talked to the Astro people about orbit determination resources.   All of the existing orbit determination software systems ran on large mainframe computers.   When I was working at Goddard Space Flight Center, the largest computer made by IBM, the 360-95, was used for orbit determinations.   RCA Astro had contracted with University Computing Company (UCC) in Dallas, TX to provided time on a Univac 1108 computer operating under EXEC8 to run a version of the Air Force Space 360 orbit software.   The 1108 supported multiprocessing and could run 3 separate programs simultaneously.   During launch operations a Univac Remote Job Entry (RJE) terminal would be installed in Vernon Valley connected to UCC by a 4800 Baud synchronous modem.   This allowed rapid input of punched card data and provided a high speed printer After launch operations were over, access to the UCC 1108 for on-orbit operations would by an LA 36 DEC Writer terminal connected via a 300 Baud dial-up modem[1].

---

1   At this point, Ma Bell would not allow direct connection of customer modems to phone lines so after the number was dialed the telephone handset had to be pushed into two rubber cups on the modem.

I just hoped that the system worked well because that would be a slow speed data link if you had to do a lot computer reruns.

After a week working with Ray, he left for California to take over the TT&C West job. I was gaining a lot of technical information from the Glōbcom engineers at Princeton Junction and the engineers at Astro, however, I was not learning much about on-orbit communication spacecraft operations. The reason for this was that no one in Glōbcom or Astro had ever done this before! Glōbcom had experience designing, building and operating satellite earth station communication facilities -- not operating communication satellites. Astro had a group of outstanding people designing and building different types of spacecraft but, with exception of the TIROS weather satellite, very little experience operating them. And the operations experience that they had was primarily with spacecraft in low earth orbit -- quite different from operating geosynchronous birds. With our tight schedule I was going to have to find people that had been flying geo birds -- and there weren't many of those!

I was told that the Satcom Programming and Control Handbook (PCH) would contain everything we needed to know about operating the spacecraft. Perusal of the draft copies of the 3-inch thick handbook revealed that it was chock full of technical data about the spacecraft. However, it contained very little of the step-by-step procedures required by people using the TT&C hardware and software to monitor, command and maneuver an operational spacecraft carrying revenue producing traffic. It appeared that if we were to have those procedures, my people would have to write them -- one more thing to do. And it was not a trivial task because it was not just writing the procedures but also building the spacecraft command lists and designing the required telemetry monitor and strip chart displays

Finally after a few weeks in Princeton Junction, my boss Gerry Long told me that the Vernon Valley building was sufficiently advanced that I could go to Vernon. I needed to be there to start staffing up the Spacecraft Operations organization.   There was still no habitable space available in the Vernon Valley Earth Station so I was pretty much working out of my house in Highland Lakes, a few miles from the station.   I was conducting job interviews at the picnic table in my back yard.   Every day I would spend some time at the station checking on progress.   Don Pidgeon was the RCA Glōbcom project engineer on the job. He was living in the nearby Playboy Hotel and shared desk space in the construction contractor's office trailer. I had known Don when I was RCA Technical Support Manager at GSFC and Don was the RCA Station Engineer at the GSFC Tracking Station at Rosman, North Carolina.   Don was a "good ole boy" and what we called a "dirty hands engineer" -- he could get things done!   Don's exploits were legion.   He was once dispatched to the NASA Tracking Station in Tananarive on the island of Madagascar to fix a failed 40-foot dish antenna.   It turned out the the mylar window on the feed had blown out and there were no spares on site.   Don went into the market in town and purchased a plastic pissoir bucket which he installed on the feed with duct tape. It put the antenna back on the air for a couple of weeks until the mylar window arrived. Another time an 1100-foot long, buried, dry air-pressurized, 2-inch diameter copper coax cable at Rosman was perforated by a lightening strike.   The air pressure unit in the station showed no pressure - the lightening had made a hole in the copper coax. Faced with the prospect of having to dig up the whole 1100-feet to find the leak, Don connected a tank of high pressure nitrogen in place of the air pressure unit.   A couple days later he walked the cable route and found the place where the leaking nitrogen had turned the grass greener. I was sure happy to have Don working with us.

With such a short time until launch, I had to find experienced people and fast.   Some of these people had to have experience operating geostationary birds because Astro was really light in that area. I hired Dave Whelan from Western Union as an Orbital Analyst.   Dave had Bachelor's and Master's Degrees in Astronomy, had previously worked at the Naval Observatory and had been in charge of orbit determination for Western Union for two years. I was also able to transfer Bob Youngblood, who was working for RCA at Goddard

Spaceflight Center (GSFC) to Glōbcom as Spacecraft Analyst.   Bob been an engineer for years on the Applications Technology Satellite (ATS) program; first at the NASA tracking station in Rosman, North Carolina and later at the ATS Control Center at GSFC.   Bob had great experience and he was also one of the smartest and most technically inquisitive persons I have ever known.

As I mentioned earlier, Ray Balon who had been the RCA Nimbus Control Center Manager at GSFC was my TT&C West Manager at South Mountain.   Duane McMillen who had been one of my Network Support Team Shift Supervisors at GSFC accepted the job of TT&C East Manager at Vernon Valley.   Besides recent experience in spacecraft operations, Ray and Duane both had many years experience in operating and maintaining satellite control hardware and software systems.

Each of the TT&C facilities required 5 Spacecraft Controllers to cover a 3-shift, 4-crew, 24-hour a day operation.   With 5 people, we could cover 24-hours a day seven days a week and have an extra person on days Monday through Friday.   The extra person could also cover about 50% of the vacation and illness absences -- the rest would be covered with overtime.   It was the most cost effective way to run round-the-clock operations.   Although these were basically technician level jobs, we were able to convince our labor relations people that they were "super-techs" with special skills and thus should not be required to be in the union with the communication technicians.   Duane was able to lure away some experienced controllers for Vernon Valley from Western Union and to transfer in a couple RCA/GSFC people.   One of the Western Union controllers that we hired was a laid-off Eastern Air Lines 727 pilot.   He made a great controller but unfortunately 8-months later Eastern called him back to work. Ray Balon was not able to get any people with controller experience but found some good supertechs who with training should work out well.[2]

---

[2] In August 1981, President Ronald Reagan fired striking Air Traffic Controllers.   At that time I was looking for Spacecraft Controllers and hired several fired Air Traffic Controllers -- they worked out great!

Each TT&C also required a Computer Technician (a union job). At that time most mini-computer users contracted with the computer manufacturer for maintenance support -- when something breaks you just pick up the phone. That works well if you only need support Monday through Friday during business hours. Contracts for computer maintenance for 7-day, 24-hour support is very expensive. I determined that it was more cost effective to have our own trained HP computer tech and to buy spares for the most critical parts. We also bought the special maintenance kits ($40,000) required for the disc drives of that era. Other spare parts were available from HP on overnight delivery and we had system redundancy by the fact that we had two identical TT&C facilities. I also wrote a Purchase Order to HP so that if we got in big trouble we could call on them for emergency help - we only had to do that on a few rare occasions. The system worked out very well, primarily because the HP equipment was reliable but also because we had good computer techs and technically qualified TT&C Managers.

We were fully manned when Astro ran the training course for Spacecraft Controllers, Analysts and Managers in October. It was the first time that we had the whole team together. We had brought the West Coast people east for the training. The first few days of the course were conducted at Astro and consisted of general system overview and an opportunity to see the spacecraft hardware in the white rooms where F-1 was nearing completion. The training then moved to Vernon Valley. The students were housed in the Playboy Hotel that also provided classroom space. Controller training exercises were conducted on the TT&C System at Vernon Valley using the Spacecraft Simulator.

As the training progressed, I became more confident that we would be able to pull this off. Having the experience of our Western Union and GSFC veterans really paid off. They added a lot to the training discussions and provided some "real world" input to such things as maneuver planning and maneuvers, thermal management and power management. Having the whole team together gave us an opportunity to discuss spacecraft operations policy and strategies and some more mundane things such as shift operations.   It was a very productive few weeks.   Only two of the team did not work out.   A South Mountain Controller had trouble with the training and despite substantial extra help at night couldn't pass and was let go.   One of Duane's Western Union recruits turned out to be an on-the-job pot smoker.   He was quickly replaced with another Western Union controller.   We were manned, trained and ready for the first launch rehearsal in early November.   The rehearsals were run by Astro, primarily for their own launch team of about 40 engineers, analysts and orbit people.   Our Spacecraft Controllers at South Mountain and Vernon Valley were the console operators and worked under Astro direction.   There were two rehearsals; one for a week four weeks before launch and another for the week immediately before launch.

I told my team that the rehearsals were a great opportunity to learn more about the spacecraft and the TT&C system.   During rehearsal weeks all of the Astro engineers for attitude control, thermal, power, propulsion and orbital dynamics would be in Vernon Valley and we should use the opportunity to pick their brains -- and we did.   Using that knowledge we started to put together rough drafts of procedures, determined required strip chart and monitor displays and designed command lists.

In the fall of 1975, the Federal Communications Commission ruled that because RCA Glōbcom owned International Communication Facilities, it could not also own US Domestic Communication Facilities.   As a result RCA spun off the domestic satellite business and formed RCA American Communications Inc. -- more generally known as RCA Americom. Phil Schneider, Glōbcom VP of Engineering was promoted to President RCA American Communications Inc.

Chapter 6 - F-1 - Learning To Fly

I have made reference a number of times to "flying the spacecraft". How do you fly something that is over 22,000 miles away and even with a good telescope only looks like a bright star? It turns out that the spacecraft talks to you -- well not exactly. It actually transmits a continuous stream of data, called telemetry, that reports the status of about 1000 data points in the spacecraft to the Telemetry, Tracking and Command (TT&C) System on the ground. Every 2 seconds the telemetry reports on such critical things as Battery Voltages, Attitude Errors, Load Bus Current and Communications Status. Every minute or so it reports on less critical or slower changing points such as Temperatures and Communications Switch Status. This continuous stream of data, transmitted by the spacecraft, is received by the TT&C Antenna, processed in the Data computer and displayed on Video Monitors and strip chart recorders near the TT&C Console. Reports of all of the telemetry values can also be printed. The Data Computer also continuously compares the incoming telemetry data against preset limits in a database and sounds alarms for data that is outside of normal limits.

So then how do you talk back to the spacecraft? How do you fix problems reported by the telemetry? The TT&C has a facility to transmit commands to the spacecraft through the TT&C Antenna. The commands are initiated by the Spacecraft Controller at the Control Computer keyboard. Most functions require sending more than one command so these groups of commands are stored in the Control Computer as Command List files. The first command in every list is a Spacecraft Clear that clears out the command register in the spacecraft and sets all values to zero. The second command in each list is a No Operation (NOOP) command that turns on the spacecraft Command Decoder so that it is ready to receive and decode the following commands. The rest of the commands are then sent at about eight second intervals. The spacecraft "echoes" each command back down to the TT&C and the computer verifies that the command is correct before executing it by sending a NOOP command. The last command in the list is a Spacecraft Clear.

Another prime TT&C function is "stationkeeping" which basically means keeping the spacecraft where it is supposed to be. The Federal Communications Commission assigns each communications satellite a Latitude (North/South) and Longitude (East/West) "Box" that is about 25 miles on a side. Natural forces such as the earth and the moon's gravity and solar wind pressure cause the spacecraft to move and change direction. The spacecraft must be maneuvered periodically to keep it within the bounds of the box. So how do we know where the spacecraft is? The TT&C ranging system transmits a series of audio tones to the spacecraft and measures how long they take to return and the phase shift of the received tones. The Control Computer takes the ranging data and calculates rather precisely the actual distance (range) to the spacecraft. In addition to the tone-derived ranging data, each range file contains azimuth (Left/Right) and elevation (Up/Down) pointing data from the Tracking Antenna. The TT&C stations at Vernon Valley and South Mountain range on the spacecraft once each hour for at least 24-hours (usually 30) and sends these range files to the offline Univac computer at UCC in Dallas.

The Orbital Analyst then runs programs on the Univac computer to determine the exact location of the spacecraft and which way it is moving. The computer processes the 30 hours worth of range and elevation and azimuth data from Vernon Valley and South Mountain and removes bad data (outliers) and smooths the data points. The smoothed data is then used to calculate the exact location of the spacecraft at the time of the first range file (epoch time). It also calculates spacecraft motion and propagates the spacecraft trajectory forward for about three weeks (an ephemeris). This information is used by the analysts to plan a spacecraft maneuver where thrusters are fired to change the spacecraft's direction of movement to keep it within the "box". As a service to our customers, so that they could accurately point their antennas, I set up an 800 number with a recording that told customers when the spacecraft would be in the "center-of-the-box".

My primary concern as we were getting organized, was how to put the people into this complex hardware and software system. Our mission was to operate the spacecraft so that we provided our customers the highest quality and most reliable communications possible. At the same time I wanted the TT&C system to be as user friendly as possible to prevent confusion, reduce operator errors and to minimize workload and stress. Spacecraft Controllers have a stressful enough job being responsible for operating spacecraft worth hundreds of millions of dollars. I wanted the total system including operator interfaces, training, documentation, analysts and management to support the controllers in the most effective ways possible.

When I was at Goddard Space Flight Center (GSFC), I had done the first-ever study of operator errors in the worldwide Space Tracking and Data Acquisition Network (STADAN). One of my responsibilities at GSFC was the STADAN Network Support Team (NST). The team kept track of everything that affected spacecraft operations in the 21 station network including any problems with the 50 satellites that the stations were supporting. The stations were required to send teletype reports to the NST anytime they had problems of any kind. The NST also coordinated with Mission Operations that monitored every satellite pass over every station to insure that we did not miss anything. We kept track of equipment failures, scheduling errors, logistics problems, software failures, prediction errors, documentation problems, spacecraft anomalies, environmental disturbances and operator errors. The NST Shift Supervisor gave a network status briefing every morning at 8 o'clock to NASA and contractor management to review all of the problems in the previous 24 hours and discuss and implement solutions. We called this "near-real-time" support and it turned out to be very effective.

I was concerned about the number of operator errors that were being reported to the NST and started a database to try to determine if the errors were random or repeated and why. As a result we were able to cure some chronic errors by equipment modifications, software changes and changes to documentation. This study had made me very aware of the kinds of things that cause people to make mistakes. I vowed to try to build a Spacecraft Operations organization that performed reliable, repeatable, operations independent of who was doing them. It wasn't easy. The Glōbcom engineers that were responsible for building the TT&C Systems thought that they knew more about operations than we the operators did although none of them had ever operated anything.

I knew that I was in trouble the first day I arrived at Princeton Junction. The TT&C Engineering Manager came up to me and asked "How do you want the Solar Array Angle data displayed?" The solar array made one rotation a day so I said "zero to 360-degrees". He said "we can't do that". I knew then that we were in trouble. It turned out that in order to get 0.1 degree resolution, the 8-bit telemetry coming down from the spacecraft could only count from zero to 45 degrees (one octant). When I suggested that if at system initialization, you told the software which octant the array was in and had the software count octants every time the angle went through zero the program would know where the real zero was and could figure it all out. In other words if the telemetry read 30-degrees and it had counted four zeros the displayed telemetry should read 120-degrees (4 x 30 = 120). They contended that was too much work and they couldn't do it.

Bob Youngblood, a very inventive guy, made a 24-hour clock that had a picture of the top (North) face of the spacecraft on the face of the clock. The clock also had two hour hands; one red and one black. The black hand showed the hour of the day at the spacecraft's longitude (local time) and the red hand showed where the sun was shining in relation to the spacecraft. We made one clock for each spacecraft and installed them in longitudinal sequence on a wall where the controllers could easily see them. If our telemetry couldn't tell us where the sun was shining on the spacecraft (and where the solar array should be), our visual aid could. Without a doubt those clocks saved our butts on more than one occasion.

As we prepared for our first rehearsal, we found out there was much to be done. Astro had a contractual responsibility to provide a TT&C System that was complete and ready to start operating Satcom 1. Not only wasn't the software complete, but none of the F-1 Command Lists required for launch operations no less those needed for on-station operations had yet been built -- literally hundreds of lists Astro asked if our Spacecraft Controllers could help out by building them. I was a little chagrined that they weren't done, because we were only a week away from the first launch rehearsal, but at that point there was not much else to do but buckle down and get the job done. We were all concerned about the launch and were trying to help Astro in any way that we could. Here it was November and the launch was only a few weeks away. We were all putting in long hours trying the get all the tasks done and checking everything at least twice.

After launch, the spacecraft normally makes 7 orbits around the earth before it is maneuvered into a geostationary orbit. At the 7th Apogee (highest point) when the spacecraft is about 22,000 miles above the earth, the Apogee Kick Motor (AKM) in the spacecraft is fired. This rocket burns 600 pounds of solid fuel in about 26 seconds which circularizes the orbit and puts the spacecraft into geostationary orbit over the equator. In order to track the spacecraft during the early orbits when it is on the other side of the earth and out of sight of the Vernon Valley and South Mountain stations, Astro leased facilities in two Intelsat communication earth stations. One of these was in Fucino, Italy not far from Rome and the other was in Carnarvon on the west coast of Australia about 600 miles north of Perth. The Astro TT&C facilities in Carnarvon were connected by a 4800 baud synchronous data line to the South Mountain TT&C. The Fucino TT&C was similarly connected to the Vernon Valley TT&C. South Mountain and Vernon Valley were also interconnected by 4800 baud data lines. There was also a 4800 baud data line from Hanger AE at Kennedy Space Center (KSC).

The first launch rehearsal was rocky but we got through it OK and learned a lot in the process. Astro had supplied a hardware spacecraft "simulator" to use for training and tests. Basically it simulated command reception and verification, provided two telemetry streams and simulated ranging and attitude data collection. We practiced ranging, attitude data collection and various commanded operations such as Spin Precession Maneuvers, Apogee Kick Motor (AKM) Fire, Dual Spin Turn, Solar Array Deployment, etc. Simulated Range Files and Attitude Data files were collected from the Vernon Valley (VV) and South Mountain (SM) TT&C Systems and sent to the offline computer at UCC for processing. Arrangements were made with NASA and Telesat Canada to obtain real spacecraft ranging files and these were successfully processed in the UCC computer. Procedures were revised, command lists were modified and telemetry displays were redesigned. Some procedures were repeated many times -- until we finally got them right.

Network tests were also conducted and telemetry data transmission and file transfer procedures checked between VV, SM, Carnarvon and Fucino. During the times that the spacecraft was in sight of either Fucino or Carnarvon, their telemetry data was forwarded to the other stations so that we could all monitor the spacecraft. Using simulator data, we verified that capability. We also verified that we could receive telemetry data from the RCA Spacecraft Checkout Station at Hanger AE at the Kennedy Space Center (KSC). Once Satcom 1 was mated to the Delta launch vehicle and powered up, the RCA people at KSC would forward live spacecraft telemetry into our launch network. The full-time "shout down" voice circuits between all the stations were tested and verified. Having had considerable experience at Goddard with problems on launch voice circuits, I wrote and issued a procedure containing calling conventions, discipline instructions and number and alphabet pronunciation guides.

We also verified the operation of the power systems at both Vernon Valley and South Mountain. Because of the critical nature of spacecraft operations and satellite communications, we needed ultra reliable electric power. Each station had two 250,000 Watt Caterpillar diesel generators. The station power system was divided into two parts a Critical Bus powering all the TT&C equipment, critical communication equipment and emergency lighting and a Utility Bus powering everything else. If commercial power failed, both diesel generators started -- they were normally up to speed in less than 15 seconds. The first generator up to speed took the Critical Bus loads. If the generator on the Critical Bus failed, the other generator dropped its loads and assumed the Critical Bus. Because TT&C critical equipment couldn't tolerate even a 15 second interruption, all of that equipment was operated from a battery powered Uninterruptible Power System (UPS) powered from the Critical Bus. The batteries could supply power for about 15 minutes, if required.

Preparations continued through November with only a short break for Thanksgiving. We continued drilling the Vernon Valley and South Mountain Spacecraft Controllers on all of the critical launch procedures. We worked on emergency procedures for the ground system. Controllers were drilled on sending commands manually from the Command and Range Tone Generator (CRTG) in case the Control Computer failed. Controllers practiced rebooting the computers -- not an easy procedure in an HP-2100 mini-computer. Reboot instructions had to be entered manually into the register buttons on the front panel and executed one at a time. Reloading software involved setting up registers manually from the front panel and then reading in the basic system from a very long, 1-inch wide, punched paper tape.

Meanwhile work continued to assure that the ground system was performing as it should. Ground system engineers conducted calibrations of the ground system hardware to identify range errors caused by phase shifts of the ranging tones in individual hardware components. Compensation for these errors were entered into the ground system Control Computers. Engineers calibrated the 13 Meter Datron Tracking Antennas at both stations by tracking the radio noise emitted by the Cassiopeia A star nebulae. The antenna azimuth and elevation shaft encoders were then adjusted to remove the errors. The System Analysts entered orbital elements into the UCC offline computer to generate antenna prediction files to drive the tracking antennas as they would during transfer orbit.

Before the final launch rehearsal Spacecraft Operations went to full operational status. Both

Vernon Valley and South Mountain TT&C Facilities were manned 24 hours a day. Spacecraft Controllers were working our regular 3 shift operation which meant that there was one controller on at each station for all three shifts (8AM to 4PM, 4PM to midnight and midnight to 8AM) with an extra person on days Monday thru Friday. For launch rehearsals and the actual launch we went to two 12 hour shifts with two persons on each shift and an extra person on days. Once the spacecraft was stable at geosynchronous orbit, we would go back to normal manning.

The final launch rehearsal started one week before launch and ended on the scheduled launch day. The rehearsals went pretty well. There were some problems with data communications with Fucino and Carnarvon, but that was fairly normal. Remember, in 1975 to communicate at 4800 baud required a 4-wire C2 conditioned telephone circuit and VCR-sized Rixon T208 Synchronous Modems that were proprietary to AT&T so we couldn't even touch them. Our technicians spent a fair amount of time working with AT&T and the local phone company to keep the circuits up.

Launch was scheduled for Thursday December 11. Count proceeded normally until someone spotted "foreign material" on one of the Castor strap-on solid rockets. The count was held while someone went out to check -- turned out to be bird poop -- big bird! The countdown was finally stopped when the final FCC launch approval was not received. Everyone was disappointed but we were convinced that tomorrow would be a better day.

Launch day, December 12, 1975 dawned overcast and cold with temperatures in the mid 30's. When I got to work that morning, I looked across the valley at the ski slopes at the Vernon Valley and Great Gorge Ski Areas. There was snow on the slopes but they had stopped making snow because of rising temperatures. Things were quiet in the station. The night shift was still on duty just doing occasional simulated range file or attitude file collection. At shift change one of the guys on the off-going shift said "I sure hope that we have a spacecraft to fly when we come back tonight".   I called South Mountain to check their status.   All systems were operating normally and the rehearsals had gone well.   There were some Astro personnel at South Mountain prepared to take over if Vernon Valley had a major TT&C failure.   We had also arranged for contractor support at both stations including HR Computer techs, antenna and RF specialists and power system specialists among others.

In the late morning we started getting Satcom 1 (F-1 once in flight) telemetry from the Cape. The RCA Spacecraft Checkout Station in Hanger AE had an antenna pointed directly at the shroud enclosing the spacecraft at the top of the launch vehicle.   The spacecraft was receiving electrical power from a connection to ground facilities.   It was very exciting -- it was the first time that we had seen live telemetry from the spacecraft.   We could see that all three batteries were being charged and spacecraft temperatures were higher than normal.   That was to be expected of course because even with cooling air being supplied to the shroud, the spacecraft was a lot warmer than it would be in space.   We wanted to launch with the batteries fully charged because the spacecraft would be spinning at 60 RPM with only a small portion of the total solar array exposed to the sun.

At 95 minutes before launch, KSC meteorologists released a weather balloon that they tracked by radar to determine the winds aloft.   High winds at upper altitudes could push the launch vehicle off course.   Winds aloft turned out to be OK so we were cleared for final countdown.

Delta 3914 carrying RCA Satcom 1 lifting off from KSC December 12, 1975 at 6:56PM EST - *NASA Photo*

The launch widow opened at 6:44PM EST and closed 12 minutes later at 6:56.   Joe Napoli, the Glōbcom spacecraft power engineer, suggested a launch pool with 12 of us each drawing one minute.   We all threw money into the pool and I drew the last minute in the launch window -- there was no way that I could win. When the countdown gets to T-20 minutes there is a built in 20 minute hold.   Another winds aloft balloon is released and the launch team is again polled for flight readiness.   Everything was OK and the count picked up. At T-5 Satcom 1 is switched to internal power -- it is then running on its own batteries.   At T-4 there is a built in 10 minute hold. As we were waiting, the Range Safety Officer reported that the launch was on indefinite hold because of a sailboat in one of the danger areas down range.   The Coast Guard was dispatched to chase the sailboat.   We passed the end of the scheduled 10 minutes and were still on hold.   Finally the Coast Guard reported that the sailboat was clear.   The Range Safety Officer cleared the launch at 6:52, we lifted off at 6:56 and guess who won the launch pool!   I have always had a fond place in my heart for that sailboat.

The Delta 3914 carrying Satcom 1 left the Cape with an earth shaking roar and an enormous cloud of smoke and steam.   It flew southeastward climbing in altitude.   As the nine Castor IV solids burned out they fell away into the ocean.   As the main engine Thor stage shut down, it also dropped away and fell into the sea.   The Delta second stage engine (a modified Lunar Module Descent Engine) ignited and continued to push higher.   After passing through the maximum aerodynamic pressure point and out of the atmosphere, the shroud split apart and fell away leaving the spacecraft in the open.   The third stage with Satcom 1 attached was now in a 100 mile circular orbit and coasting.   Passing over the equator (first nodal crossing) the third stage STAR 37E engine ignited pushing the spacecraft into what is known as a geocentric transfer orbit (GTO).   When the third stage burn was complete, it released the spacecraft which was spinning at 60 RPM.   The high point (apogee) of the orbit was at 19,350 nautical miles above the earth, which is the synchronous altitude, and the lowest point (perigee) was about 100 miles above the earth.   The plane of the orbit was inclined to the equator about 28.5 degrees which is the latitude of Cape Canaveral.   In about three days, when the Apogee Kick Motor (AKM) is fired on the 7th apogee, it will circularize the orbit at synchronous altitude and take out the 28.5 degree inclination so that it will be right over the equator at all times.   The Delta third stage is in a similar orbit, however, because the perigee is so low the third stage is slowed down by atmospheric drag every time it goes through perigee.   The orbit gradually decays and in a few months the third stage will burn up in the earths atmosphere.

The spacecraft was now on its own traveling across Africa.   Its track took it over Namibia and South Africa and over the Indian Ocean.   The Astro team at Carnarvon Australia was waiting to pick up the radio signal from F-1.   The Carnarvon antenna was being driven along F-1's predicted path based on orbital elements supplied by the KSC range tracking data.   The spacecraft was well above the optical horizon at Carnarvon before its radio beacon is received. That is because the spacecraft was pointed so that its antenna beam was not optimized on the earth. Carnarvon's first task was to do a Spin Precession maneuver to get the spacecraft antenna beam pointed at the earth.

At Vernon Valley, I had been nervously waiting for the first telemetry data from Carnarvon. We all breathed a big sigh of relief when we started to get Carnarvon data -- F-1 looked great. The next three days were very busy at all the stations. We took range files and attitude files and sent them to UCC for processing. Spin Precession maneuvers were performed to start to put F-1 in the correct attitude for AKM fire. As the spacecraft went through each perigee, atmospheric drag slowed the spacecraft and changed its attitude requiring more attitude data collection and Spin Precession maneuvers. The F-1 initial orbital elements were based on launch vehicle tracking data. As we took more and more range files a more definitive orbit was obtained and was used for AKM fire planning. The spacecraft controllers were performing well and operations were going pretty smoothly. Lou Muhlfelder, Astro's chief attitude engineer and one of the designers of the Satcom spacecraft kept a plot of all of the attitude data taken. In the lower left corner of the plot were two concentric circles. The inner circle defined the range of the most optimum attitude for AKM fire. The outer circle defined the maximum range that he was willing to accept for AKM fire. As the days went on the attitude data points moved down toward the lower left corner. We were getting there.

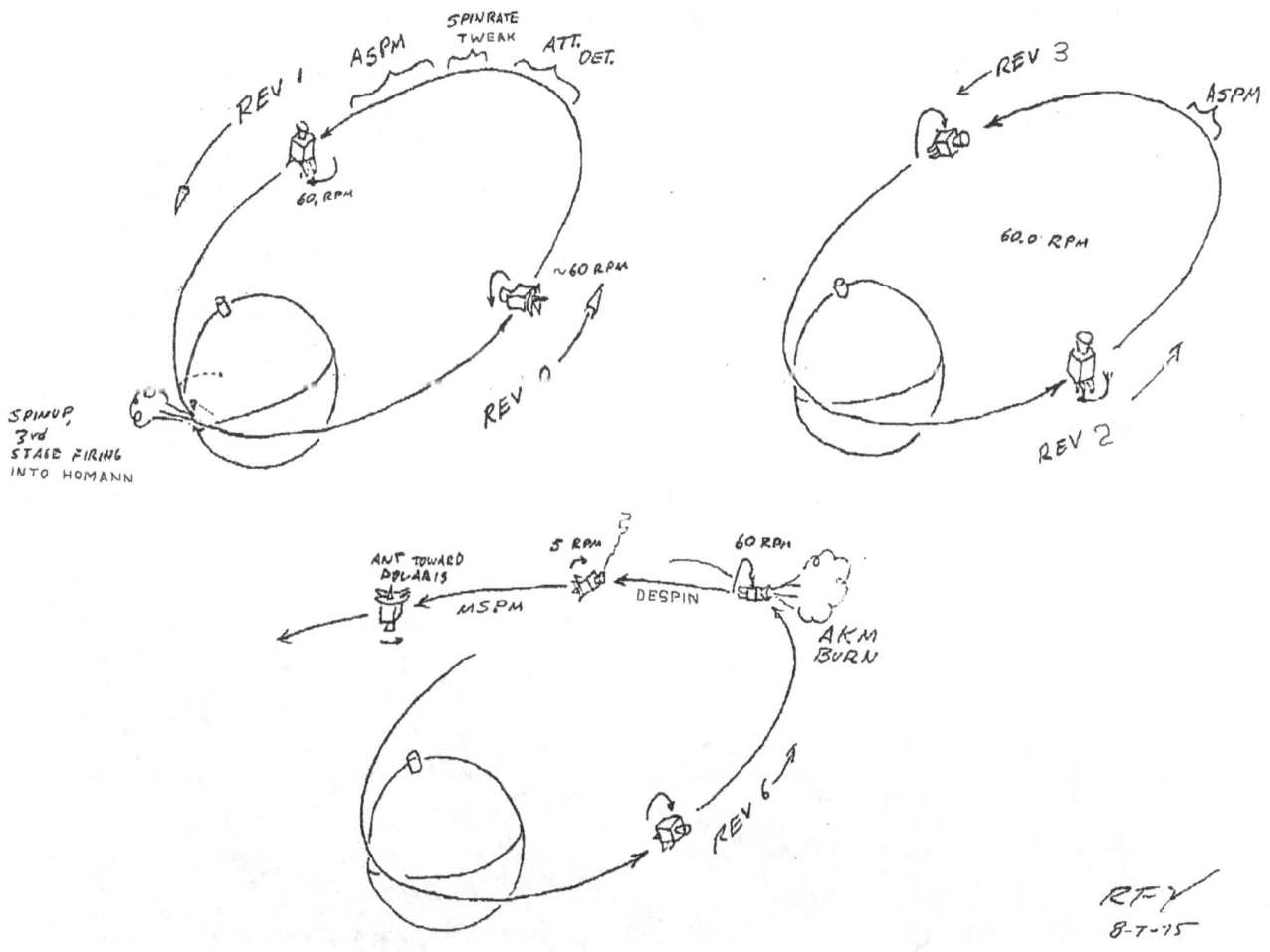

This is one of Bob Youngblood's training aid "cartoons". It shows the spacecraft trajectory and attitude from the Delta 3rd Stage burn through AKM fire, despin and orientation for the dual spin turn. He also shows where we did Attitude Determinations (ATT. DET.) and Spin Precession Maneuvers (ASPM) - *Robert F. Youngblood*

Within RCA the AKM Fire event was getting to be a really big deal. It was scheduled for about 8PM on Monday December 15. The guest list at Vernon Valley kept getting larger. In addition to our own immediate management we were also expecting Andy Conrad, President of RCA Corp., Howard Hawkins, President of RCA Glōbcom Inc. and Phil Schneider President of Americom. Things were crowded enough with just the working stiffs there and now we had VIP's. We set up a strip chart recorder (SCR) in the back of the operations room to display AKM Pressure during firing. Some chairs were arranged in front of the SCR for our VIP guests. It got very quiet as we approached fire time. The controllers at Vernon Valley and South Mountain were running the AKM Fire procedure in parallel and bringing up the

73

command lists together.   The Vernon Valley controller would actually send the Enable and Fire command lists unless they were unable to -- in that case South Mountain would take over. Fortunately everything ran smoothly.   VV sent the AKM Fire Enable list that powered up the firing circuits and the Enable status was verified in telemetry.   At the designated time the AKM FIRE list was released.   Because these were hazardous commands, the controller also had to turn and hold the Hazardous Command Key.   The commands went out and almost immediately the pen started to move on the VIP's SCR showing the pressure rising.   In 26 seconds it was all over to a round of applause and cheering.   The Tracking Antenna operator reported that the antenna had stopped moving because the spacecraft had stopped -- exactly as expected.

The AKM Burn Was Successful!

L to R - Facing to camera - Andy Conrad, Pres. RCA Corp., the author, Howard Hawkins, Pres. RCA Glōbcom Inc. Back to camera - Jack Lewin, Mgr TT&C Engineering RCA Americom, Dennis Elliot, VP Finance RCA Americom, Phil Schneider President RCA Americom - *RCA Americom Photo*

At that point the VIP's departed for the Playboy Hotel to continue the celebration while all of us settled down for a very busy night.   Although the spacecraft was stopped at synchronous altitude, it was still spinning at 60 RPM on the wrong axis and the solar arrays were folded up. There was really no rush -- the spacecraft was in a stable mode and the folded array was supplying enough power for the bird's modest needs at this time.   Nevertheless, everybody wanted to get the F-1 deployed on station so that testing could begin.   Ranging was started immediately.   We needed to get an orbit to find out how effective the AKM burn had been -- particularly in bringing down the inclination.   If we didn't get a good burn we would have to use on board fuel to get the spacecraft on station thus shortening its life.   As it turned out we were OK in inclination and were drifting at a nice pace toward our orbital location at 135 degrees West Longitude.   Now comes the interesting part -- we have to slow down the spacecraft spin, get it reoriented along orbit normal, execute the dual spin turn, deploy the folded up solar array and further despin the spacecraft so that it can lock up on the earth.

Thruster firings were commenced to slow the spacecraft spin rate.   The objective was to slow the spin rate from 60 RPM down to 5 RPM.   Data from the horizon sensor is used to measure spin rate.   When the spacecraft had slowed to 5RPM thrusters were fired in a Spin Precession Maneuver to position the spacecraft parallel to orbit normal.   In this position the spacecraft antenna is pointed North and its spin axis is parallel to the earths spin axis.   Attitude Data files were taken to confirm that the spacecraft was correctly positioned for the Dual Spin Turn.

The Dual Spin Turn is accomplished by turning on the spacecraft Momentum Wheel Assembly (MWA) that spins at 6000 RPM while the spacecraft itself is spinning at 5 RPM.   Because the wheel spin axis and the spacecraft spin axis are 90 degrees apart there is a gyroscopic interaction of forces that starts the spacecraft to cone about its spin axis.   Over a couple of hours, the size of the coning keeps increasing until finally the spacecraft is spinning about the MWA axis.   Now the spacecraft is in its desired attitude spinning slowly.

The solar array and its support arm have been folded up against the North and South faces of the spacecraft.   The hinges in each of the folds are spring loaded and would unfold unless restrained.   When the array is folded against the panel, "belly band" cables hold the arrays down.   The array is released by firing pyrotechnic cable cutters.   The explosive gases are constrained within the cutter to prevent contamination of spacecraft optics.   Redundant cutters assure that the cables will be released.   The design of the cables and its attachments assures that the cables will be thrown free of the spacecraft.

Deploying the solar array is another one of those "hold your breath " moments.   Command list Arm Pyrotechnic Bus is sent to apply power to the firing circuits.   When all is ready, the command list "Fire SA Pyros" is sent.   There is no telemetry to show success, however, as the array deploys into the sun the Solar Array Current increases.   The spacecraft is still spinning slowly so as the array comes about and points directly at the sun, the amount of Solar Array Current will indicate if the array is completely deployed.   Fortunately for F-1, the array deployed exactly as expected.

Chapter 7 - Making History

Tuesday, December 16, 1975 was a historic day. When the RCA Americom F-1 spacecraft arrived on-station in geosynchronous orbit, it became the worlds first commercial three axis stabilized satellite. It contained 24 communication channels, each capable of carrying a color TV signal or 972 telephone circuits. That was more than twice as many channels as any previous communication satellite. And unlike earlier power-limited spinning spacecraft, all of the channels could be operated 24-hours a day for 8-years. It was designed to make a lot of money for its owners -- and it did.

By contrast, two earlier three axis stabilized communication satellites were experimental vehicles with specialized, limited communications capability. The US Government NASA ATS-6 satellite with a 30-foot communications dish was the first three axis stabilized communications satellite, however, it was only used for communications experiments. The Franco-German government sponsored Symphonie three axis spacecraft had only two communications channels and they could not be operated through most of the twice-a-year 40-day eclipse seasons.

Being first of course has some drawbacks. We in Americom Spacecraft Operations got to find out how you fly this new spacecraft of revolutionary design with many previously untested or unproven features. How well does magnetic attitude control work at synchronous altitude? What effect do solar flares and magnetic storms have on spacecraft attitude? What happens to spacecraft attitude when the plume from North face thrusters bounces off the solar array? How well does the spacecraft thermal design handle the 24 hour a day solar radiation cycle? And we were not alone. Even the Astro folks who designed the spacecraft were feeling their way. This became apparent a few days after F-1 was on station.

As soon as the spacecraft was stabilized, spacecraft bus testing started. These tests were to prove that all of the components of spacecraft bus (everything but the Communications Subsystem) were operable, including redundant units. The tests also established a performance baseline for the spacecraft that would be used for reference throughout its life. RCA Americom had planned a gala party at the nearby Playboy Hotel to celebrate the succesful launch of F-1. On the morning of the party, Astro was testing the Attitude Control Subsystem. I wasn't directly involved in the tests so don't know exactly what happened. I looked out my office window that overlooked the TT&C operations room and noticed there seemed to be some excitement among the Astro engineers. Duane McMillen, the Vernon Valley TT&C Manager came into my office and said "they've lost the spacecraft". "What do you mean", I said, "I can see telemetry updating on the monitors". "Well" he said "evidently F-1 has lost lock on the earth". As I went into the ops room, I could see Lou Muhlfelder, one of the designers (and patent holders) of the attitude control system. He had our balsa wood model of the satellite in his left hand and his right hand fingers were spread to show the spin and momentum vectors. He was explaining to some of the engineers what was going on. There seemed to be a consensus among the engineers that the problem was that the pitch control loop was open so the spacecraft was stopped while turned away from the earth. Closing the pitch loop should resolve the problem, however, there was some reluctance to send the required commands. Bob Miller, Astro's Satcom 1 Program Manager, said that he didn't want to do anything until the dynamics group did a detailed analysis. John Christopher, Americom's Spacecraft Engineering Director wanted to send the Pitch Loop Close commands, however, Bob was adamant that he wasn't doing anything until the situation was completely analyzed. With everyone involved with this problem, the Launch Party was postponed until the following night. On the following day the analysis was complete, the pitch loop was closed and F-1 turned around and locked up on the earth. As much as all the Astro people knew about the design and construction of this bird, they were being very conservative in their approach to problems because it was all new territory to them also. Having solved that problem, we all went to the Playboy and partied.

In retrospect, I believe that I know what happened to the Astro team that caused the crisis. We had been told by Astro that if we lost lock on the earth, we should send a command list called HELP! We had put that advice in the draft procedures that we were assembling. A few months post launch, after the Astro folks had left Vernon Valley, F-1 lost lock on the earth in the middle of the night because of an unexpected Earth Sensor scan inhibit caused by the sun. In accordance with our procedure the Astro-prescribed command list HELP! was sent. About 15 minutes later when I got into the station, I found that F-1 was still pointed away from the earth and didn't seem to want to come back. Bob Youngblood, Dave Whelan and I couldn't figure what was wrong -- we followed the procedure and nothing happened. Finally at about 3 in the morning, we called Jack Frohbieter, the F-1 Mission Director, and he conferenced in Bob Cenker one of the attitude engineers. We told them what we had done and after some thought Bob Cenker said "doesn't the HELP list have an Open Pitch Loop command?" We quickly checked and sure enough he was right. We quickly sent the Close Pitch Loop command and good old F-1 came right around and locked up on the earth. I am sure that was what happened to Astro months earlier -- they lost pitch lock, sent the HELP list and ended up thinking they lost the spacecraft. Unfortunately for us they didn't correct the list and that lesson needed to be learned twice.

F-1 was assigned the orbital slot of 135 degrees West Longitude on the equator and due south of the Yukon Territory. F-2, to be launched in about 3 months was assigned to 119 degrees West due south of Los Angeles. F-2 was designated to carry the Alaskan traffic. Americom, however, wished to get the Alaskan traffic off of Westar 2. F-1 was positioned at 119 West temporarily so that as soon as it was checked out, the Alaska traffic could be transferred from Westar. Once F-2 was launched in late March, it would be checked out at 128 degrees West (an unused slot) and then flown to 119 West to take over the Alaskan traffic. F-1 would then be flown to 135 West.

Spacecraft bus testing continued rather uneventfully until we got to maneuvers. The East/West (longitude) maneuvers went well using all of the possible thruster combinations and using both Attitude Control Electronics (ACE) 1 and 2. Then came the real test, North/South maneuvers. Astro knew that the exhaust plumes from the North face thrusters (1, 2, 3 & 4) impinging on the solar array and could cause serious attitude disturbances. We new that the approximate position would be where the solar array was basically bisecting the four thrusters. The exact optimum position would be where the plume forces on one side of

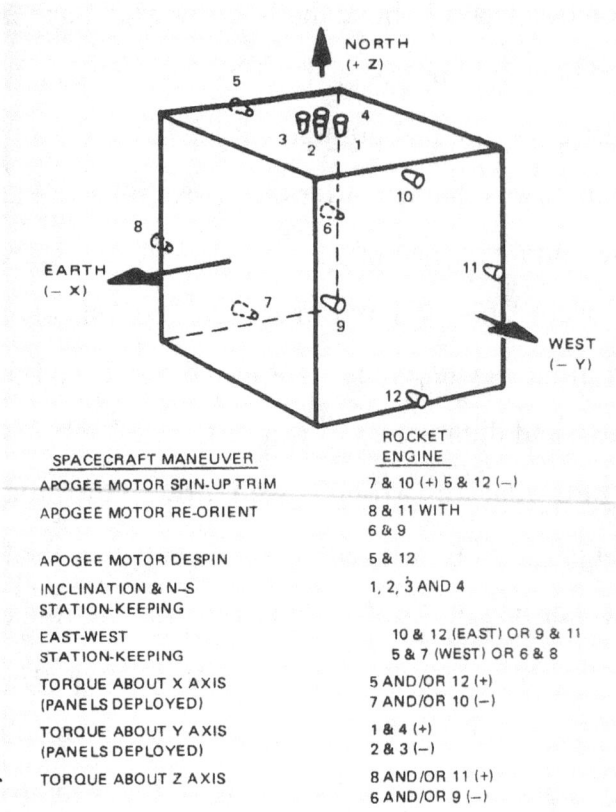

SPACECRAFT MANEUVER	ROCKET ENGINE
APOGEE MOTOR SPIN-UP TRIM	7 & 10 (+) 5 & 12 (−)
APOGEE MOTOR RE-ORIENT	8 & 11 WITH 6 & 9
APOGEE MOTOR DESPIN	5 & 12
INCLINATION & N–S STATION-KEEPING	1, 2, 3 AND 4
EAST-WEST STATION-KEEPING	10 & 12 (EAST) OR 9 & 11 5 & 7 (WEST) OR 6 & 8
TORQUE ABOUT X AXIS (PANELS DEPLOYED)	5 AND/OR 12 (+) 7 AND/OR 10 (−)
TORQUE ABOUT Y AXIS (PANELS DEPLOYED)	1 & 4 (+) 2 & 3 (−)
TORQUE ABOUT Z AXIS	8 AND/OR 11 (+) 6 AND/OR 9 (−)

the array equalled the plume forces on the opposite side. Astro had advised us that they would perform tests to identify the solar array position that resulted in the minimum disturbance and that could be controlled by the attitude control system -- the so called "sweet spot". They performed plume tests by firing various combinations of north face thrusters while varying the solar array position. After substantial data was gathered, they reported that the "sweet spot" had been determined. We were told that the defined "sweet spot" was the array angle that we should use for all North/South maneuvers. They then proceeded to do a short N/S maneuver (we didn't really need one at that point) using the Attitude Control Electronics (ACE) 1 and announced success. We duly noted the designated solar array angle and put it in our draft procedures.

A number of days later Astro scheduled another short N/S maneuver to test the operation of ACE 2. The solar array was slewed to the designated "sweet spot" angle and stopped. At the scheduled maneuver start time, the north face thrusters started firing. Almost immediately the spacecraft started to roll severely and the ACE automatically aborted the maneuver because of excessive attitude errors.   After the thrusters stopped firing the spacecraft was nutating plus or minus a degree in roll.   If we had been carrying communications traffic our phone would have been ringing off the hook.   The Astro attitude engineers had to fire thrusters to remove the nutation.   Obviously chagrined at this turn of events, the Astro engineers retired to the analysts room to review the data and to try to figure out what happened.   The next day they did more test firings of the north face thrusters while changing array angles.

It turned out that there was no "sweet spot".   A North South maneuver on F-1 generally required about 20 minutes of of North face thruster burn.   The center time of a North/South maneuver is at the time of Ascending Node, that is when the spacecraft is crossing the equator northbound.   That time changes by 4 minutes a day -- so the maneuver time changes each day -- which also means that the angle of the sun on the array changes each day.   The light weight solar array panels bend slightly with temperature -- when the sun angle changes, the shape of the array changes -- when the shape changes, the attitude disturbance torques change!   So now what do we do?   New procedure -- the day before a scheduled N/S maneuver, we now have to perform the Astro plume test at the center time of the next days maneuver to determine the optimum array position.   It turns out that the optimum angle changes with time of day, day of year and from year to year.

Using North face thrusters, the time for North/South maneuvers is in the middle of the afternoon during the summer and in the middle of the night during the winter.   North/South maneuvers were too complex to be done by Spacecraft Controllers alone.   So this meant that we would all have to trudge through the snow to the station two nights in a row to do a North/South in the winter --Thanks a lot.   The range of the optimum solar array angle turned out to be quite small, and a little hard to predict.   This resulted in some exciting maneuvers with attitude disturbances closely approaching the spacecraft abort limits and more than one ground-commanded abort.   Despite these problems, however, we were able to operate the spacecraft safely and provide highly reliable service to our customers.

The only other serious problem discovered during bus testing was that one Command Logic Decoder (CLD) did not work.   Although disappointing, it was not considered serious because F-1 had four CLD's (two per Command Receiver).

The Communication Subsystem is literally the payload on a communication spacecraft.   The 24 communications channels are the product we sell to our customers.   It is imperative that each channel provide the highest quality transmission.   Each channel is a transponder that acts like a "bent pipe" repeater -- that is it takes a signal from the ground, turns it around and sends it back.   A TV signal, for instance, is transmitted from an earth station on the ground (South Mountain for example) received in the spacecraft, amplified and then transmitted back to the portion of the earth covered by the spacecraft antenna pattern.   In the case of F-1, the antenna pattern covered all 50 US States plus central and western Canada and the northern Caribbean.

The unit in the transponder that amplifies the signal is called a Traveling Wave Tube Amplifier (TWTA).   In the case of F-1 the TWTA transmitted 5 watts -- about as much as an old CB radio.   RCA Astro bought the TWTA's from Hughes Electronics; one of the few companies manufacturing flight-rated TWTA's at that time.   The TWTA consists of two parts the actual Traveling Wave Tube (TWT) itself and the power supply that provides several low and high voltages to the tube.   Because the TWT requires power voltages of over 10,000 volts, it is very important that the power supply be manufactured with extreme care. Care must be taken to round off sharp points where arcs may form and assure that solder joints do not have air filled voids in them.   The whole power supply is encapsulated in a plastic potting material that must be applied carefully to prevent air filled voids in the material.   It turned out that Hughes, because of high demand for flight-rated TWTA's at that time, subcontracted the manufacture of the power supplies to another company with no previous experience building flight rated power supplies.

When the spacecraft bus testing was complete, Astro started to turn on the payload. When an individual TWTA is commanded on, power is applied to the TWTA filament and about 2-1/2 minutes later, after the tube has warmed up, the high voltage is turned on. Because of the importance of transponder operation to the companies revenue, a "color bars test" was performed immediately after turn on. South Mountain uplinked a carrier to the transponder, established transponder saturated power and then modulated the carrier with a standard TV Color Bar test pattern. If a satisfactory color bar picture was received in Vernon Valley the transponder was considered operable -- empirical tests to fully characterize the transponder would come later. All of the transponders came on and passed the "color bars" test.

Americom then started the communications system tests to fully characterize the performance of each channel. These tests require a suite of carefully calibrated test equipment at an earth station. They also require calibrated radio frequency paths to and from the antenna used for the tests. Initial testing of F-1 was conducted from Vernon Valley, however, communication testing of later spacecraft was conducted from South Mountain. Many tests were conducted on each transponder including the alternate paths through the Backup Receivers. Parameters measured included the amount of uplink power required to saturate the transponder, the signal level received from the satellite, the frequency response of the transponder and the amount of intermodulation distortion when multiple carriers are used. F-1 provided 24 channels by reusing the 500 Megahertz of bandwidth usually used for 12 channels; once in the horizontal polarization and once again in the vertical polarization. Tests were conducted on each channel to measure the isolation between the signals in the two polarizations. The results of all of these tests and many others proved that the Satcom spacecraft would provide high performance transmission of television, single and multiple carrier telephone and data traffic and the small carriers in Single Channel Per Carrier (SCPC) transmissions.

Unlike later spacecraft, F-1 did not have any spare TWTA's -- if a TWTA failed the channel was lost. During the F-1 communications testing, Transponder 22 turned off spontaneously. We commanded it back on and within a short time it turned off again. It continued to fail and we continued to turn it back on. This turned out to be a big mistake. Earlier we mentioned that Hughes had used a new subcontractor to build the TWTA power supplies. Analysis showed that the TWTA probably turned off because of "outgassing". In the manufacture of flight rated power supplies great care has to be taken to insure that solder joints and plastic potting compound do not contain voids that could contain pockets of air. Evidently the TWTA power supply contractor did not use processes designed to prevent voids. As a result, after the power supply was in the vacuum of space the air (and moisture and manufacturing gases) entrapped at about 15 PSI forces its way out and provides a path for high voltage arcs. As the spacecraft gets warmer the pressure of the trapped gas increases and more escapes. In hindsight, our repeated turn-ons probably caused sufficient high voltage arc-overs to burn carbon conduction paths in the potting material and thus render the power supply useless.

Later, a couple other TWTA's experienced spontaneous turn-offs. We did not try to turn them on. We noted that they occurred at the warmest time of the day which would be consistent with outgassing. We followed a procedure that allowed us to safely restore those TWTA's to normal operation. At the warmest time of the day, we commanded the TWTA on. The TWTA on-board circuits, powered the filament on and then allowed about 2-1/2 minutes for the tube to warm up before turning the the high voltage on. As the high voltage was about to turn on, we commanded the TWTA off. The procedure was designed to get the TWTA as hot as possible and force the gas out without causing a high voltage arc over. After weeks of that procedure, we would let the high voltage come on for a few minutes that made the TWTA even warmer and then command it off. Over weeks we gradually increased the high voltage on time until we left the TWTA on -- and they generally stayed on. One of them subsequently turned off at the warmest time of the day of the warmest day of the year. That confirmed the diagnosis of outgassing -- the higher temperature forced out the last bit of gas. It stayed on after that.

Through all of this we had worked on our procedures and command lists and basically took over operation of F-1 under Astro's watchful eye. It would not officially be Americom's bird until they legally accepted it from Astro and put traffic on it. As noted earlier F-1 was located at 119 degrees West in what would ultimately be F-2's slot. Americom wanted to get the Alaskan traffic and HBO off Westar and on to F-1. After F-2 was launched and checked out at 128 Degrees West Longitude, it would be flown to 119 West and we would do a flyby transition. F-1 would then be moved to 135 West.

A transition plan was set up to repoint the Alaskan, San Francisco and Valley Forge earth station antennas from Westar 2 at 123 degrees West Longitude to RCA Americom F-1 at 119 degrees West Longitude. Initially Alascom would have 6 transponders in F-1 primarily carrying telephone traffic. One of these transponders would contain hundreds of small Single Channel Per Carrier (SCPC) carriers for the Bush Telephone Service. New "lower 48" telephone carriers would be established in F-1 transponders from each of the other Americom earth stations serving New York, Chicago, Houston and Los Angeles. When scheduled, the plan would be executed over several hours in a procedure designed to minimize the length of interruptions to service. The transition of the dozens of SCPC carriers would take a few days because technicians had to journey to the bush stations to move the antennas.

Typical Alaskan Bush Earth Station antenna. Note the very low elevation angle because the station is so far North. The station was often installed next to the bush village General Store and the store owner designated as an agent for Alaskan Telephone. Villagers could make phone calls from the single phone booth and pay the store owner - *RCA Alascom Photo*

In mid February 1976, F-1 was declared operational. Americom officially took over F-1 and the Astro people went home. My organization, Spacecraft Operations (SPACEOPS), had been preparing for this moment for many months. I had established responsibilities within SPACEOPS and defined rules for notification and escalation in the event of spacecraft anomalous performance or other emergencies. I had proposed and gotten approved by Spacecraft Engineering a policy on the spacecraft normal operating configuration, ground rules for handling spacecraft failures and a failure reporting system. The FCC had defined an operating "box" of plus or minus 0.1 degree of latitude and longitude around the spacecraft's assigned location. As a matter of policy, SPACEOPS would operate in a plus/minus 0.08 box to insure better service for our customers and to preclude accidently exceeding the 0.1 box. We established a Center of the Box phone number with a recorded message containing info on when the spacecraft would be in the center of the box so that customers could point their antennas precisely at the satellite. We had developed, reviewed and approved procedures for operation of all of the spacecraft major systems. We also set up a system of Operations Bulletins (OpsBulls) to cover temporary or emergency procedures. Every day we issued a Daily Operations Plan (DOP) listing times for all spacecraft events for the next 24 hours including maneuvers, ranging, eclipses, sun or moon interference in Earth Sensors and special tests. The DOP also listed the Duty Analyst to be called in case of a problem. The objective was to insure that operations were documented, scheduled and repeatable no matter who was on shift night or day.

The Communications Transition Plan was executed, all of the traffic was transferred and F-1 was finally carrying revenue bearing communications. As we took over F-1 we were also in the midst of preparations for the F-2 launch which was scheduled for March 25. During the final rehearsal and launch operations, F-1 would be flown from South Mountain. Vernon Valley and South Mountain complimented one another, Every week since F-1 was on station, we had been shifting prime operational responsibility for F-1 back and forth between Vernon Valley and South Mountain. For the F-2 launch I sent Bob Youngblood and Dave Whelan to South Mountain to support F-1 operations and to provide ongoing training for the South Mountain Spacecraft Controllers.

Chapter 8 - The Hanger Queen

Satcom 2 was about two months behind Satcom 1 in the construction flow at Astro. Although long lead time items for Satcom 3 had been ordered, construction on that bird would not start for some time. There was not quite the schedule pressure on F-2 as there had been for F-1. During the press to complete F-1 to make its launch date, failed units on F-1 were replaced with tested units destined for F-2 and the F-1 units sent for rework. After rework and testing, the former F-1 units went into the flow for F-2. Effectively F-2 was the "hanger queen" for F-1.

Having just been through the launch campaign for F-1, both the Astro and Americom teams were ready to do it again for F-2. There were a number of meetings to review the F-1 experience to determine what should be changed to improve operations for F-2. Procedures and command lists were revised and timelines updated. TT&C tests were conducted at Vernon Valley and South Mountain using F-1 live telemetry data and F-2 data from the simulator to assure that the system could in fact support two spacecraft. We were still finding problems with the TT&C software so the Astro programmers were providing regular updates. Because we were now operating a real spacecraft, we didn't want to chance new problems being introduced in software revisions. New software builds were first installed in South Mountain and run for successfully for about 10 days before installing in Vernon Valley. We did not want to take a chance that both stations could become inoperable because of software failures.

The first F-2 launch rehearsal went relatively smoothly. The primary problems were with the "high speed" data links between Vernon Valley (VV) and South Mountain (SM) and the data links to Fucino, Italy and Carnarvon, Australia. High speed in 1976 was 4800 Baud and required a Rixon T208A modem that was about the size of an old VCR. These modems could not operate on a normal telephone circuit. They had to have a 4-wire (one pair transmit -- one pair receive) telephone circuit that was "C-conditioned". The conditioning was provided by

AT&T by adding inductors or capacitors to the circuit in their telephone offices to improve the frequency response and phase delays.   The main difficulty was that for the long distances involved, the circuits went through many telephone offices and there were many opportunities for troubles -- and we had them.   It seemed like almost every day we were turning at least one of the circuits back to AT&T with troubles.

As we started the final rehearsal, Dave and Bob left for South Mountain and South Mountain took over primary control of F-1.   Even though Vernon Valley was the primary station for the launch, it also backed up South Mountain on F-1.   Vernon was routinely scheduled for ranging on F-1.   F-2 events were added to our Daily Operating Plan (DOP) and we held coordinating meetings with the Astro Mission Director to schedule all operations for both spacecraft at both stations.   We had planned F-1 maneuvers and subsequent ranging so that they occurred before and after the F-2 launch and AKM fire window.

As usual for Delta, the F-2 launch was scheduled for Thursday March 25.   Unfortunately high winds aloft caused the launch to be scrubbed on Thursday. The Delta count picked up again Friday morning for a scheduled 5:47PM liftoff from Launch Complex LC-17A at Cape Canaveral.   The launch itself was uneventful and F-2 was picked up as expected as it came within range of Carnarvon, Australia.

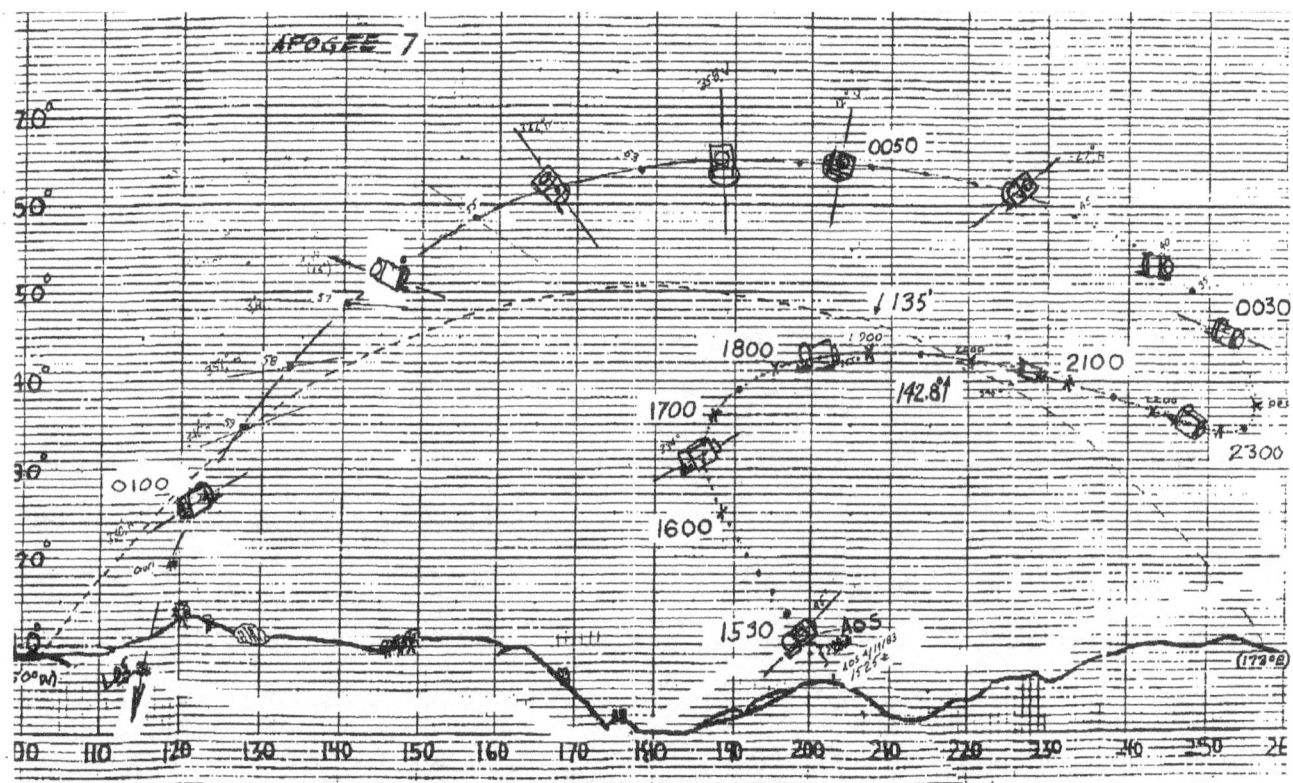

This is one of Bob Youngblood's plots of the expected spacecraft trajectory as seen looking South from South Mountain. Azimuth angles are along the bottom and elevation angles on the left side. The spacecraft track is annotated with the GMT times. Note the mountains that form the local horizon. - *Robert F. Youngblood*

Bob Youngblood, who was out at South Mountain, was an amateur astronomer. He had brought with him a war surplus US Air Force Bubble Sextant. These sextants were used from the domed plexiglass port on the top of World War II vintage aircraft to measure the angle from the horizon to celestial bodies. Because it is difficult to see the horizon from aircraft at altitude, the sextant used a bubble level to provide the horizon reference. Bob had figured out that on the day after launch F-2 would pass over South Mountain close to perigee altitude and shortly after sunset -- the ideal time to spot an object in space. He and Dave Whelan and TT&C Manager Ray Balon and other South Mountain personnel went outside away from the lights of the station as the time approached. Bob used the sextant to locate some reference stars to determine where F-2 would appear and sure enough -- there it was. It appeared as a bright star moving rapidly across the sky. Although South Mountain was in the dark, F-2 was still in the sunlight and that gold mylar thermal blanket certainly reflected a lot of light.

91

The next couple of days were very busy with F-2 ranging, Spin Precession Maneuvers and Attitude Data Collection.   All of these activities were focussed on assuring that the spacecraft was oriented in exactly the correct attitude for the AKM firing on Sunday.   The AKM was fired successfully, the despin and Dual Spin Turn executed, the solar array deployed and the spacecraft locked on the earth.   Everyone was checking telemetry and looking at temperatures and voltages and attitude data and F-2 appeared to be in good shape -- until Bob Youngblood called to Vernon from South Mountain on the shout down circuit.   "The F-2 solar array has stopped moving" he said.   Someone acknowledged the call but didn't question him further.   Bob called again "The F-2 solar array has stopped moving and the solar array current is dropping".   That finally got some Astro attention at Vernon Valley.   They queried Bob and he said "Its been stopped for about 20-minutes -- the drive motor is pulsing but the telemetry Solar Array Drive (SAD) angle hasn't changed".   At Vernon they commanded the SAD to Stop and then commanded it into Reverse.   The SAD angle started to change in the reverse direction as the array moved away from where it had stopped.   After a few minutes they commanded SAD Stop and then SAD Normal Forward and it started to move forward -- until it got to the same place it had stopped before. It stopped again.   The drive seemed to be working OK so something must be blocking the array or its drive shaft. We were all sure that the SAD drive motor was OK but Astro commanded the backup drive motor on and had the same problem. There was real concern by all the engineers and management of both Astro and Americom about F-2's health and prognosis. If the solar array couldn't follow the sun, it couldn't power the payload.

By now the sun angle had changed enough that the array output had dropped sharply. Bob Youngblood recommended that they slew the array in reverse to get around to the other side of whatever was stopping the array, where the sun would ultimately catch up. Commands were sent to put the SAD in Slew Reverse so that it would move relatively rapidly around to the other side of the obstruction. No one wanted to go too close to whatever it was that stopped the array because of the possibility of getting stuck and not being able to back out. The array was stopped at a point where it was just beginning to get sunlight again. It would be left there until the sun was full on the array and then put in Normal Forward so it would follow the sun for the next 20 hours or so. The time of arrival at the danger area was calculated and telemetry alarms set to warn the controller to stop the array before it was stopped by the obstruction. Command lists can be set to execute at a specified time. The SADSTOP command list was set up to execute automatically at the same time as the alarm. No one wanted to take any chances that the array could get jammed and not be able to move at all.

By now, Astro engineers were looking at the spacecraft documentation to see it they could identify possible culprits for the problem. It didn't take too long. The North and South solar arrays are at either end of a long shaft that goes through the box structure of the spacecraft body. The shaft is actually two shafts joined together by a round coupling. The coupling has a protruding weldment seam along one side. It appeared that a nearby semi-rigid, corrugated, coaxial cable going from the Telemetry Beacon Transmitter to its antenna had popped loose from a mounting clip and was leaning against the coupling. When the seam came around on its daily rotation, it caught in one of the cable corrugations and the array stopped. That doesn't sound like much of an impediment, however, the solar array drive couldn't push very hard. In the weightlessness of space it doesn't take much to rotate the array so a small low-torque stepping motor was used to drive it. The array made one rotation per day so that the drive was essentially a clock motor and it could easily be stalled. The cable probably came out of the clip during the heavy vibration of launch and now it was laying up against the shaft coupling..

In the forward direction, we knew where the obstruction was because he had hit it. Now knowing what was causing the problem, the engineers could calculate where the obstruction probably was on the other side. They defined the bounds of the danger zone in both array angle and time of day. The solution to the problem was to stop the array when it came to the beginning of the danger area. The array remained stopped until the sun had moved enough so that there was no longer enough solar power to support the spacecraft loads. As the batteries started to discharge, the array was slewed in reverse until, after about 40-minutes, it came to the other side of the danger zone and it was again stopped.   It remained stopped as the sun came around until the sun was square on the array and its electrical output was at maximum. The solar array drive was then commanded into Normal Forward mode where it remained all day until it again approached the danger zone when it was once again stopped.   This was repeated every night at about 7PM Eastern time.   Initially the entire REWIND sequence, as it came to be known, was done with a single command list that had to be initiated by the Spacecraft Controller.   Later REWIND was implemented in a software solution that did not require any operator intervention.   The REWIND command lists were still retained for backup.

The only problem with all of this was that we had to move the array off the sun to do North/South maneuvers (Remember the plume problem?).   In order to do this, we had to suspend the software program.   We also had to be sure that when we finished the maneuver, we got the array back to exactly where the software thought it should be at that time of the day. If we didn't get it right, the array wouldn't get stopped at the correct time and might run into the cable obstruction.   Most times, to be safe, we had the Spacecraft Controllers do the REWIND manually with command lists after the array had been moved for a maneuver..

We were also having a more general problem with both F-1 and F-2 with thermal control. The temperatures in the spacecraft were considerably warmer than those predicted prelaunch. We were particularly concerned about how hot the batteries were getting and how the higher than normal temperatures would affect their life.   By design, excess power from the solar array is dissipated in shunt resistors mounted on the central cylinder in the body of the spacecraft.   This heat was contributing to the overall higher temperatures of spacecraft components.   The solar array on both F-1 and F-2 was offset slightly off the sun so that the array was only producing the power actually needed by the spacecraft loads plus a little bit extra.   In this condition, only a small amount of excess power was being dissipated as heat in the spacecraft.   Doing that was no problem at all for F-1, however, for F-2 this offset had to be compensated for in the timing of the REWIND command lists and software.

We were very careful to keep news of the F-2 solar array problem from the general public. The REWIND system was working and there was no impact on spacecraft performance.   We were afraid that misinformation about the problem would cause both our existing customers and potential customers to question F-2 reliability.   And we thought that our secret was safe in the far reaches of space -- WRONG!   One day Bob Youngblood got a call from a colleague at the MIT Millstone facility in Massachusetts. In addition to the Haystack Precision Radar, the facility also included some large telescopes.   The caller said "I have been looking at your F-2, and although I can't make out the details of the spacecraft, I notice that at about 7 o'clock at night the reflected light changes."   Bob explained to him what was going on and asked him to be discreet in talking about it, which he was.

Ultimately though, rumors started to appear in some of the trade press about a solar array problem -- in F-1!   Of course, because the cable traffic was on F-1 with literally hundreds of antennas pointed at it, there was much more interest in it than in F-2.   Bob Cooper, a "TV-DX" and "backyard dish" enthusiast and writer had been leading the charge in writing about the "F-1 array problem".   When he became editor of the Community Antenna Television Journal, he wrote a series of articles about supposed Satcom problems.   In order to set the rumors to rest and get the facts out, RCA invited Cooper to visit Vernon Valley.   Cooper also had a TV program so RCA arranged for an NBC crew to tape his visit and interviews.   We used that occasion to reveal that the problem was on F-2 not F-1 and to demonstrate that it was causing us no operational problems.   It worked out to be a public relations success and quieted down the talk about Satcom problems.   It also showed RCA management that we needed to be much more open and forthcoming about any problems we might have.

For 2-1/2 years we followed the REWIND procedure every night.   One night, while the SAD was in Reverse Slew, the software failed to send the STOP command and the solar array plowed right through the Danger Zone from the reverse side -- the cable didn't stop it!   We were still reluctant to probe the Danger Zone because of the possibility that the array could get stuck and not be able to move at all.   A few months later, a Spacecraft Controller was doing the REWIND manually, made an error and again slewed the array in reverse through the Danger Zone.   The next day after reviewing the Solar Array Drive Motor Current history, it did not appear that there was any drag on the array as it went through the Danger Zone.   That night, we left the SAD in Normal Forward and it proceeded normally through the Danger Zone. Evidently the first time the software slewed the array through in reverse, it knocked the cable out of the way.   From that day until its last day of service F-2's 24-hour array rotation operated normally.

Unfortunately, what we didn't know at the time was that the solar array itself was already damaged by the REWIND procedure.   Every time the array was slewed around during REWIND, it underwent thermal stresses similar to those experienced by the array assembly during eclipse. As the array rotated and the sun angle changed, the array flexed and bent slightly because of expansion and contractions of the metal materials in the array structure. At the same time, because the materials in the cells and their substrate are different, differential expansion takes place during the rapid temperature changes (approx. 100 degrees C in a few minutes) causing stress on the individual cell electrical connections.   To minimize the power loss from individual cell failures, the cells are connected in series/parallel strings.   The number of cells connected in series varies depending on voltage requirements -- normally about 6. Typically 3 or 4 strings are connected in parallel for reliability.   Because of the mechanical stresses of REWIND cell strings began to fail reducing the solar array power output

Geosynchronous spacecraft normally experience 80 eclipses a year or 720 over a 9-year life. By contrast, F-2 experienced 912 REWIND eclipses in the first 2.5 years of life plus the 720 real eclipses over its 9-year life for a total of 1632 eclipse-like events!   The array was thermally stressed almost twice as much as designers had anticipated. The effect was that we started to see drops in array output as individual cell strings failed. At the end of life, F-2's solar array output had dropped so much that it could no longer provide power for its total communication payload.

While we were consumed with the problems of the F-2 Solar Array, the spacecraft had otherwise been performing flawlessly. All of the F-2 spacecraft bus tests and communication tests had been completed at its temporary location of 128 Degrees West Longitude. It was now ready to be placed into operation as the primary Alascom spacecraft. F-1 was currently carrying the Alascom traffic in its temporary location at 119 Degrees West. The plan was that we would fly F-2 eastward from 128 West to 119 West. The two communication receivers (one for horizontal polarization and one for vertical polarization) were turned off so that F-2 communications were inoperable. An East/West maneuver was executed to start F-2 drifting eastward in late May 1976. When a synchronous spacecraft is moving eastward, it is below synchronous altitude; when moving westward it is above synchronous. As F-2 approached, we maneuvered F-1 so that it was at the eastern edge of the 119 West "box". As F-2 entered the western edge of the box, we executed an East/West maneuver on F-1 to start it moving westward. As F-1 passed above F-2 in approximately the center of the box on June 5, 1976, F-2's communications receivers were turned on and then F-1's were turned off. Except for a few seconds of multipath interference when both F-1 and F-2 receivers were on, the transition from F-1 to F-2 was made without serious interruption to traffic. As F-1 continued westward, an East/West maneuver was executed to stop F-2 in the 119 box. F-2 was now carrying 6 transponders of Alaskan traffic, Americom telephone traffic and HBO.

F-1 was assigned to an orbital location of 135-degrees West Longitude. It had been drifted over there from 119-degrees and was now on-station and operating normally. Although we did not yet have much PLC traffic, message carriers had been established in F-1 from all of the Americom earth stations including those serving New York, Los Angeles, San Francisco,

Philadelphia, Chicago and Houston. HBO had a TV test carrier up in Transponder 20. One day I got a call from someone in the Operations Control Branch at Goddard Space Flight Center. They published the Satellite Situation Report that listed all objects in orbit. He said "Did you know that there is a military communications bird located at 135-West?" "No I didn't", I replied, "How the hell can they do that?" He said he didn't know but he thought that it was a US Air Force communications spacecraft and he gave me a phone number to call. I called the number and it turned out to be an Air Force facility. The 2nd Lieutenant I talked to was rather reluctant to tell me anything and would not identify the spacecraft. I told him that RCA's F-1 communication spacecraft was located at 135 plus or minus 0.1 degrees in East/West and North/South and asked him if that concerned him? He did not seem to know much so I asked to speak to his superior officer. After some time on hold a different guy picked up but he would still not identify the spacecraft or himself. I asked him if he could tell me what kind of stationkeeping they were doing. He put me on hold again and in a couple of minutes someone else said that the spacecraft was plus or minus 0.5 degrees in East/West and plus or minus 6.0 degrees in North South -- this thing was sweeping out a huge amount of air space! I asked him if he was concerned that our two spacecraft might collide. He said that he would pass that along to his chain of command. Well if the Air Force wasn't concerned, I sure was! I discussed it with Bob Youngblood and Dave Whelan. They were both very concerned. Bob pointed out that because of the complex motion of both spacecraft, it wasn't like two dots approaching one another, it was more like two electric fans passing through one another. That certainly wasn't a very comforting thought.

In the 1970's that type of orbit was common for government spacecraft. We did very tight stationkeeping, that required a lot of fuel, so that out commercial customers could just leave their antennas in one position. The military had expensive tracking antennas that could follow their birds as they moved around -- they would much rather use the weight on the spacecraft for their spook payload instead of for fuel. They did no North/South maneuvers at all -- the spacecraft was delivered on station with 6-degree inclination. Solar pressure changes inclination at about 0.85 degrees per year so in 7 years it will be at zero degrees inclination and in another 7 years it will be at 6 degrees on the other side. It only took a small amount of fuel for their not-very-often East/West maneuvers.

At this point in 1976 there were not many spacecraft at synchronous altitude -- so why did we end up sharing our assigned slot with a military spook who wouldn't talk to us. I recommended to our management that we move a little way further west to 135.8 degrees West Longitude to stay out of their way. We stayed at 135.8 until the spring of 1978 when more spacecraft were being launched and the FCC told us we had to go back to 135. We ultimately found out that our "spook" neighbor was Synchronous Meteorological Satellite 2 (SMS-2) a Hughes-built weather spacecraft supposedly belonging to NASA. If that was so, why did I end up talking to 2nd Lieutenants who wouldn't answer my questions about it -- is a puzzlement!

By late 1976. things were going pretty well. Both F-1 and F-2 were running hotter than was expected but the high temperatures did not seem to be problem. Most of the South panel of the spacecraft was covered by what looked to be a mirror but was in fact a thermal radiator that reflected the suns heat and radiated spacecraft internal heat into space. The twenty-four 5-Watt TWTA's also radiated about 10 Watts of heat each into the South panel. The South panel radiator was needed to get this heat from the TWTA's out of the spacecraft. Smaller radiators were located on other panels near hot components to keep them cool. It was ultimately determined that residue from the AKM rocket firing landed on and coated these radiators reducing their efficiency thus causing the spacecraft to be warmer than expected.

The F-2 REWIND procedure was now functioning in an automatic software program requiring very little intervention by the Spacecraft Controllers. Marv Freeling, one of our engineers who had previously worked in the RCA Tube Division, in Harrison, NJ, was working with our balky F-1 TWTA's coaxing them into staying on.

One morning, I was sitting in my office when I heard a telemetry alarm go off followed shortly by a controllers voice saying "All the transponders turned off!". That certainly got my attention. As I got out on the floor, I looked at the F-2 Comm display and sure enough it showed that all the TWTA's were OFF. The communications technician reported that they had lost all traffic on F-2. A quick look at the power system telemetry showed that all of the tubes could not be off. At about this time Bob Youngblood arrived in the Ops Room and said "The Receiver Power Supply has failed -- it also supplies power for the TWTA telemetry". We immediately had the Spacecraft Controller send the commands to switch in the Backup Receiver Power supply. F-2 communications traffic was immediately restored. In retrospect, that was a mistake. We had restored communications but we didn't know if the power supply had actually failed or just shut itself off. I immediately made a policy that if any component shuts off, it is commanded back on at least once before switching in a redundant unit. We always wondered if the power supply had failed or not. Years later, near F-2's end of life a communication receiver failed noisy. We took that opportunity to try the previously "failed" power supply on our noisy receiver -- it was dead! We switched in the backup receiver and restored service. I was beginning to feel that our new Spacecraft Operations organization was doing pretty well. We were operating F-1 and F-2 safely, were working our way through some of the individual spacecraft idiosyncrasies and had shown that we could respond to emergencies quickly as a team.

Chapter 9 - Building The Business

All of our efforts over the year building up to the F-1 launch had been directed toward getting two spacecraft built and launched and building the facilities necessary to support on-orbit operations. Now that F-1 and F-2 were operational, Americom needed to get the transponders filled with revenue bearing traffic and start making money with the huge investment that RCA had made in the enterprise.

Phil Schneider had been Vice President Engineering for RCA Glōbcom when the Satcom project was initiated.   When the FCC ruled that RCA had to separate its Domestic and International communications businesses, Phil was named president of the newly formed RCA Americom.   I have to say that I was a little disappointed.   I felt that the big engineering effort was behind us and that what we needed was a marketing and sales guy.   Unfortunately I was to be even further disillusioned by our new president.

In our first year in 1976, business was growing but evidently not as fast as planned. The 1976 Business Plan showed us growing the PLC business by leaps and bounds. My experience with PLC at American Satellite told me that much growth in one year was not realistic. I had heard some rumors of possible belt tightening measures but didn't think that it would affect Spacecraft Operations. Phil was visiting Vernon Valley for what I thought was just a general look at how things were running. Earlier in the morning I had given him the usual $5 dog and pony show on Spacecraft Operations.   He had asked a few questions about the birds and nothing seemed to be amiss. A little later, after the Earth Station Manager had given him the communications tour, Phil came into my office and sat down. He talked about how revenue wasn't quite as high as expected and how he felt that he needed to cut expenses. He told me that he was going to close the South Mountain TT&C Station. I was dumbfounded. It took me a few seconds to collect my thoughts, but finally I said "You can't do that." He said "Why not we have two stations?" I then proceeded to tell him that the backup protection in our system was in having two stations. Within each station there were potential single point failures, primarily the computers, and if we didn't have both stations we would not have redundancy. This appeared to be news to him. He seemed crestfallen that his big cost savings plan wasn't going to work.

After Phil had left and the more I thought about it, the more angry I got. This guy who had previously been in charge of engineering didn't even understand the reliability and maintainability design of the Spacecraft Control System. And what made it all even worse was that he obviously hadn't talked to anyone else about his plans. This didn't say very much about the decision making process in Americom. To say the least I was rather disillusioned.

Fortunately, help was on the way. In January 1977, Andrew F. Inglis was appointed Americom President and Phil Schneider left for Western Union. Andy had a long history as a Sales Engineer in the RCA Broadcast Products Division. It was said that he was on a first name basis with the Chief Engineers of every TV Station in the US. One of his first official acts was to transfer the responsibility for spacecraft transponder assignments from Engineering to Finance. He said "transponders are the only asset that we have to sell and I want to be sure that we do that in the way that will make us the most money". Way to go Andy!

The 1968 Carterfone case in which the FCC allowed equipment, other than that provided by the telephone company, to be connected to telephone lines was the first step in a series of deregulation decisions by the FCC and the courts that opened US domestic communications to innovation and competition. RCA Americom planned to use their new communication satellite capability to capture a major portion of the Private Leased Channel (PLC) telephone business away from AT&T. PLC circuits were used by business and commercial users to interconnect their various offices. These circuits were dedicated full time to the customer and allow them to pick up the phone and speak immediately to a colleague in a distant company office. Americom was not alone in competing with AT&T for this business. MCI, Southern Pacific (later SPRINT), Western Union and American Satellite were also formidable competitors. RCA was already carrying quite a bit of PLC in the F-2 Alaska traffic and a few circuits between New York and San Francisco.. Now Americom had to start building on this experience and expand it to all the markets where we had earth stations.

From the early days of satellite communications it was expected that television transmission would be a large part of any satellite company's business. Unfortunately by the 1970's, AT&T had a virtual stranglehold on TV network transmission for the three major networks (ABC, CBS and NBC). AT&T had worked with the networks from the beginning of television in the 1940's and had built its transmission network to their requirements. The networks were very comfortable with AT&T and it would be difficult for anyone else to break into it.   Cable TV, however, was a different story.   In the early 1970's, the Community Antenna Television (CATV) business in the United States was struggling. Most of the cable TV systems were in rural areas where off-the-air television reception was poor. These systems provided the three TV networks and independent TV channels from nearby cities plus a few independent channels imported from more distant cities. Cable systems in metropolitan areas were having great trouble trying to compete against over-the-air TV. What cable systems had in common was that most were deeply in debt. Building cable systems, which at that time, were 32 channels or less, required a large capital investment. The amount that could be charged to their customers was limited when most of the programming was off-the-air albeit sometimes imported by microwave links from distant markets.   Rulings by the Federal Communications Commission limited the number of distant TV stations that a cable system could import thus further limiting what they could charge subscribers. The cable companies badly needed other sources of programming to expand their subscriber base.   Many of them were experimenting with different forms of Pay TV primarily by "bicycling" motion picture films and video tapes around between cable systems.

In 1965 Charles "Chuck" Dolan, whose Sterling Television company was a supplier of cable equipment, had won the New York City cable franchise for the southern half of Manhattan island. Teleprompter won the franchise for the northern half. Dolan formed Sterling

Manhattan Cable and built the first completely underground cable system.   By 1972, Dolan was having problems because growth at Manhattan Cable, as it was now known, had stalled at about 20,000 subscribers. Time Inc. had bought 20% of Manhattan Cable and Dolan went to them with a proposal for a pay movie channel he called the Green Channel.   The new channel would offer almost-new movies and sporting events distributed by microwave links to CATV systems. The channel was renamed Home Box Office (HBO) and was transmitted on the American Television Company (ATC) microwave network to cable systems in the northeast US.   After the first year, despite "free days" and other promotions, HBO had only signed up about 8000 subscribers.   At that point Dolan sold Manhattan Cable and HBO to Time Inc. and used the money to buy the Cablevision, cable system in Long Island.   Time Inc. installed Gerry Levin as President of HBO.

When RCA won permission from the FCC in 1974 to launch Satcom 1 and 2, Sid Topol, president of cable equipment manufacturer Scientific-Atlanta, saw a great business opportunity. He already had an FCC approved design for 10-meter diameter satellite communication dishes that could be used by cable systems. He invited interested cable industry parties to a meeting at Time Inc headquarters in Manhattan.   In attendance, besides Topol, was Lou Donato from RCA Glōbcom Marketing, Gerry Levin of HBO/Time Inc., Monty Rifkin of American Television & Communications (ATC), the microwave company distributing HBO, and Henry Harris of COX Cable and Amos Hostetter of Continental Cable.   The idea Topol presented would benefit everyone.   He proposed to use RCA's satellite to distribute HBO programming, eliminating the complex of microwave stations.   Scientific-Atlanta could build the ground stations using his 10-meter dishes, and the cable systems would gain subscribers.   RCA and HBO agreed.   Rifkin thought that only 10 dish antennas would be required because he would use his ATC microwave network to distribute programming to multiple cable systems.   It soon became apparent that it would be much cheaper to put a dish at every cable headend. Bobby Rosencrans, a cable system owner in Vero Beach, Florida, saw the potential and offered his 10,000-customer cable system as a test bed for the satellite distribution scheme.

Then HBO got lucky by landing a contract, at very little expense ($50,000), to televise The Thrilla from Manila, the heavyweight championship fight between Muhammad Ali and Joe Frazier in the Philippines. With only a day or two to spare, a Scientific Atlanta satellite receiving dish was installed at the Vero Beach cable headend. Another was installed at a cable system owned by Monty Rifkin in Jackson, Mississippi and one of Sid Topol's dishes at Scientific Atlanta was also used. On 30 September 1975, a television signal took a roundabout path from Manila, first by uplinking through an Intelsat satellite to California, then by landlines and microwave relays to New York and then by RCA microwave to their Valley Forge earth station. Valley Forge uplinked the signal through RCA's leased transponder in Westar 2 to reach Florida, Mississippi and Atlanta. The Southern Cable Conference was in Atlanta at the time and Topol invited all the cable guys out to see the fight at Scientific Atlanta. The Ali-Frazier broadcast was of high quality and was a great success, and soon satellite receiver dishes were popping up at cable system headends throughout the country. Teleprompter ordered 50 dishes for their cable systems around the country. Scientific Atlanta was selling dishes as fast as they could make them - Sid Topol's vision of the future was coming true. Cable television finally had the unique service it needed to make inroads into metropolitan and suburban markets. HBO was now truly a national service.

An interesting sidelight to this story was the battle to get the FCC to allow smaller receive-only satellite dish antennas. As noted above, in 1974 the FCC required licensing of receive-only dish antennas and would not allow any that were smaller than 9-meters in diameter. These antennas cost about $100,000 installed. On October 20, 1975, the American Broadcasting Company (ABC), representing its network of radio and television stations, petitioned the FCC to retain the large dish standard. Fearing competition from cable TV systems ABC wanted to limit the cable company's access to a wider range of programming and in fact proposed even tougher dish antenna standards. On June 24, 1976, the Community Antenna Television Association (CATA), representing cable TV operators, petitioned the FCC to allow antennas as small as 4.5 Meters. These smaller antennas only cost between $30 and $40 thousand -- a huge saving that would put satellite delivered services within the reach of most CATV systems. Both petitions questioned the 1975 FCC antenna standards -- ABC thought that they were not tough enough, CATA thought that they were overly restrictive and denied satellite service to many small and medium sized communities. The FCC considered both of these petitions together and on December 15, 1976 ruled that the CATA had won -- almost immediately 4.5-meter receive-only dishes began springing up at cable headends all over the US. The cable boom was underway. In 1975 there were 3506 cable systems serving nearly 10 million subscribers -- 10 years later there were 6600 systems serving 40 million subscribers.

Once F-1 was on station in mid 1976, HBO (who had been operating temporarily in two transponders on Westar 2 leased by RCA) moved into their two F-1 transponders. HBO broadcast programming for the East Coast in transponder 24. The same programming was broadcast in transponder 20 three hours later for the West Coast. HBO was a good deal for the cable TV companies because HBO split the monthly fee with the cable operator. Initially, HBO customers paid $5 a month with the cable operator getting $2.50 and HBO $2.50. By the end of 1977, HBO was on 262 cable systems and had more than 1.6 million subscribers. For the first nine years of operation HBO was only on air nine hours a day. Programming came on at 3PM and went off at midnight. The programming originated in HBO's facilities at 1100 Avenue of the Americas in Manhattan. It was transmitted to the RCA Americom CTO at 60 Broad Street on Manhattan Cable facilities and then RCA carried it on the microwave link from 60 Broad St. to the Vernon Valley Earth Station where it was transmitted up to F-1. HBO was a stickler for video and audio quality. Americom engineers had run extensive tests on the microwave radio link and earth station uplink facilities. They had replaced compensation amplifiers that regenerate sync with straight video amplifiers and alined everything. Nevertheless, Ed Horowitz, HBO's Chief Engineer at the time, could always find a little something to gig us about when he visited. "Can't you see that piano keying" he would say. Ed was tough but fair and we got along well. We were happy to have a customer who delivered a high quality product that demonstrated our flawless satellite transmission.

Following HBO's success, Atlanta entrepreneur Ted Turner put up his Superstation WTCG on RCA's F1, Transponder 6 on December 17, 1976 thus becoming the second programmer available to cable systems all over the US. Actually Turner had been thinking seriously about satellite distribution of WTCG for some time. Turner, who owned the largest advertising company in the southeast had been worried about the impact of radio and television on his billboard business. In 1969, wanting to hedge his bets, he bought Atlanta UHF television WJRJ, which had lost $800,000 that year. Over the next couple of years Turner bought a lot of old black and white movies and showed them endlessly and changed the name of his company to Turner Communication Group and the TV station to WTCG. By 1972, WTCG broke even and the FCC had changed its regulations to allow cable systems to import distant TV signals. Since then WTCG had been picked up off-air and microwaved to various cable systems in Georgia and by 1973 it started making money to the tune of $1-million. Turner also had been buying syndicated programming for WTCG that was not governed by the FCC's syndicated exclusivity rule. This meant that his programming, including older almost ageless off-network TV series such as the Brady Bunch, Beverly Hillbillies, Bewitched, I Dream of Jeanie and Hogan's Heroes, could be distributed by satellite all over the US. Turner controlled TV rights for Atlanta pro sports teams and had built a wrestling arena in his Atlanta headquarters so that he could broadcast live wrestling matches on WTCG. The focus of the station was old movies, sports and classic sitcoms. Turner had hired Ed Taylor, a satellite expert to run his satellite operations. When the FCC told Turner he couldn't own a TV station as well as the service that sent its signal to cable providers, he created Southern Satellite Systems, and then sold it to Taylor for $1. This gift ultimately made Taylor a millionaire.

Starting on December 17, 1976 WTCG was picked up off the air by a UHF antenna on the roof of Ed Taylor's Southern Satellite Systems Atlanta earth station, fed to his F-1 satellite uplink and made available to cable systems all over the country. Unlike HBO, Turner and Taylor didn't sweat the technical details -- they just wanted to be up on the bird and making money. I occasionally looked at the WTBS multi-burst vertical interval test signal (VITS) -- it looked like the performance of a cheap home TV set -- think Muntz![3]   WTCG may not have been pretty but it worked -- folks all over the country started calling the 800 numbers to get their Popiel Fishing Gear and Gensu Knives.   Taylor charged cable companies 10 cents per month per subscriber for WTCG.   Each year Taylor billed on the basis of the subs reported on the FCC Form 325 "Annual Cable Operator Report" -- so the billing amount only changed once a year.   One cable company in Newton, Kansas only had 17 subs when he signed up in December 1976!   Turner also created Turner Broadcasting System (TBS) and began broadcasting Atlanta Braves games throughout the country via WTCG, now renamed WTBS. Later he started using the satellite to bring Braves away games back to Atlanta and also put them up on WTBS.   I believe that he was the first to use a communication satellite to bring away-games home.   The sports on WTBS made it very popular and more and more cable operators were adding it to their systems.   WTBS had become the nation's first superstation, reaching over 2 million homes by 1978.   In those early days, TV on the satellite wasn't scrambled -- if you had a C-Band dish antenna you could get the service.   Many cable operators were stealing the satellite programming and pocketing the subscriber fees themselves. But they didn't know Ted!   When listeners sent in for their Popiel Fishing gear, the delivery ZIP code was checked to see if the cable company serving that ZIP was paying for WTBS -- if they weren't, they promptly got a bill.

HBO and others used a different system to catch thieves. Television advertising agencies, in those days, paid people all over the country to watch TV and make sure that the ads that the agency had paid for are really being run on the TV at the time they were supposed to be shown. The same people were also being paid to report back what pay services were being carried on the local cable system so that the programming services could check if they were getting paid. If the local cable company wasn't paying for the service, they got a bill based on the number of subscribers they were claiming in their FCC Form 325 report.   Around 1982, HBO decided that signal theft, not only from cable systems but also from about 200,000 "backyard dishes"

---

3   In the 1950's "Madman Muntz" designed and marketed his "cool chassis" black and white TV that used only 17 tubes instead of the conventional 30 and sold for under $100.   It worked reasonably well in urban areas but performance degraded rapidly in fringe conditions.

was getting to be a big enough problem that it advertised for bids for a video scrambling system.   HBO also felt that it had to get control of its product -- at this point if a cable guy didn't pay there was nothing HBO could do -- they couldn't turn him off.   With a scrambler, of course, they could.   HBO awarded the contract for scramblers to Ma/Com for the Video Cipher system.   The results were dramatic, HBO's receivables, that had been running about 90 days dropped to about 50 days.   They also picked up about 250 Cinemax affiliates that had been trying the movie service without letting HBO know.   It also opened up a new revenue stream from backyard dish owners.   By the time that scrambling was started, the number of backyard dishes had increased to about 1.5-million.   They already had a free preview of HBO and now many of them signed up to pay for it.

The backyard dish owners were incensed, however, by the fact that the signals that they had been getting for free were now scrambled.   HBO offered subscriptions to dish owners for $12.95 a month. Dish owners felt that the price was too high because there was no competition and triggered a national movement to more strongly regulate the cable industry and force them to stop anti-competitive pricing.   One individual decided to make a more direct attack on the problem. On the morning of April 27, 1986, at 12:32 AM, John R. MacDougall, a satellite TV dealer in Ocala, FL was working at Central Florida Teleport, a company that uplinked video services to satellites.   He swung a transmit dish over to point at Galaxy 1 with the transmitter tuned for Transponder 23 -- HBO's channel.   With the transmitter at full power, he overrode HBO's programming of "The Falcon and the Snowman" for 4-1/2 minutes with color bars and the message:

The color bars and message uplinked by MacDougall as recorded at the time. The recording of the video proved to be his undoing. - *Public domain*

A number of people recorded the jamming signal.   The FCC obtained a copy of the recording for analysis.   The character generators of that era used to insert text were fairly crude and analysis of artifacts in the letters allowed the FCC to identify the make and model of the generator.   The manufacturer helped the FCC identify who had bought the generator. MacDougall was fined $5000 and put on probation.   At the time of the incident, we in the industry were quite concerned that there would be copycats interfering regularly with satellite video transmission.   Fortunately, the swift apprehension and punishment of the culprit proved a deterrent for any other would be "Captain Midnight's".

Initially, the scramblers went through a number of problems with insufficient security resulting in the Video Cipher I and then the Video Cipher II.   It worked well enough, however, that it was also adopted by Viacom and Showtime. Ma/Com was later bought by General Instrument (Jerrold) and even later by Motorola.   Ultimately digital TV distribution made the Video Cipher obsolete.

In 1961 when Andy Inglis was working in RCA Broadcast Sales he got to know the owner of a struggling religious TV station WYAH (Channel 27) in Portsmouth, VA. The station was badly in need of upgrading and Andy worked to get financing and other help to put together a deal for RCA to modernize the TV station equipment. The owner, Pat Robertson greatly appreciated Andy's efforts on his behalf. When Andy arrived at Americom he again got in touch with his friend Pat Robertson. Robertson over the years had put together a group of TV stations broadcasting as the Christian Broadcast Network (CBN). He had also constructed a CBN campus in Virginia Beach, Virginia containing not only his broadcast facilities and offices but also the newly formed CBN University. Andy convinced Pat that he should go nationwide by putting his programming up on F-1. On April 27, 1977 the Christian Broadcasting Network, broadcasting from CBN's facilities in Virginia Beach, VA became the third cable customer on F-1. Unlike HBO and WTBS, CBN was offered to cable operators free. The CBN satellite feed was not the same as those broadcast over the CBN TV stations. Although the feed did contain some broadcast content such as the "700 Club" it also contained other programming produced just for the satellite feed.

Now that there were several desirable services on F-1, more and more cable operators were installing dish antennas to downlink them for their system. Other programmers began to realize that the number of antennas now pointed at F-1 represented a lot of cable subscribers. Programmers became more interested in obtaining space on F-1 to access the growing cable market. Viacom was very interested in starting a channel to compete with HBO. In July 1, 1976 Showtime went on the air on a small cable system in Dublin, CA. showing "Celebration" a concert special featuring, Rod Stewart, Pink Floyd and ABBA. Viacom wanted to get Showtime up on F-1 to compete with HBO but did not have any facilities required to playback broadcast quality video tapes and uplink them to the satellite. Andy Inglis made Viacom an offer they couldn't refuse -- RCA will build you a video tape playback facility in the Americom Vernon Valley Earth station. RCA technicians will run it for you under your technical supervision until you can get your own facility built. Viacom signed up for two transponders on F-1 and on March 7, 1978 Showtime was up and running from Vernon Valley.

Although this was a real coup for Americom, my enthusiasm for the tape center project was limited.   In order to make space for the new Tape Center, my Spacecraft Analysts were thrown out of their office and moved into rather cramped quarters in what had been the station storeroom.   The walls of the old Analysis Office were torn down and the room enlarged. Everything else in the earth station became very cramped.   The stuff that had been in the storeroom that my analysts now occupied, was stuck in any open space that could be found around the station.

The Tape Center, when completed was pretty jammed up too.   Broadcast quality video tape at that time was two inches wide and one 16" diameter reel of tape lasted only 20 minutes. There were six huge RCA TR-600 Quadraplex Tape Machines in the Tape Center.   Three were used for the East Coast feed and three for the West Coast.   The playback heads rotated at 14,400 RPM on air lubricated bearings requiring large external air compressors in a new shed outside the building.   There was also a large film island used to broadcast programming from 16mm movie film, 35mm movie film and slides.   There was also a central control console, editing equipment and backup 3/4" Umatic cassette tape machines.   The Umatics normally ran in parallel with the TR-600's machines in case the 600 failed.

L to R - Dave Gardner, Susan Piluso, John Moran and Jacque Zembrzuski at the console for Showtime East and West. Note the backup U-Matic tape machine in the lower right hand corner. - *RCA Americom Photo*

L to R - RCA technicians Frank Ahrens and Ron Cangelosi calibrating one of the six TR-600 2-Inch Tape Machines in a rather cramped space in the Showtime Tape Center - *RCA Americom Photo*

There was also a sort of culture clash between the communications and TT&C technicians and the more "artsy craftsy" types working in the tape center. The groups would play pranks on one another. One night we were executing a North/South maneuver on F-1 and noticed that the Tape Center techs were watching a porn flick on one of the spare U-Matic tape machines. Showtime programming was going out as normal. Steve Agid, one of my Spacecraft Analysts, proposed a gag that we all thought would be pretty funny. He called in on the outside phone line to the Tape Center and in his best irate country voice said "This is John Smith in Olathe, Kansas. I was watching Showtime and all of a sudden I have this fornagraphic picture". We all watched through the Tape Center glass door. Almost immediately there were screams "Is UMatic 2 going to air? -- Is UMatic 2 going to air? What the hells going on?" Of course in a few seconds they could hear our peals of laughter and caught the joke -- they didn't think it was nearly as funny as we did.

The Tape Center techs, did however, literally have the last laugh. The RCA operated Tape Center was a temporary arrangement to fill the gap until Viacom could build its own facilities. RCA had been running the Tape Center for about two and a half years when Viacom completed their new facilities in Happhaugue, Long Island. A date was set for the switch over from the RCA Vernon Valley Tape Center. Using the editing equipment, the RCA Tape Center techs had put together a 30-second credit tape with pictures of Vernon Valley and all of the names of the RCA staff. They talked to the Showtime supervisors about rolling the tape just before the cutover. The Showtime guys said "absolutely not". The RCA people would not be denied. The cutover was scheduled for 12:30 PM. As noted before, the 2-inch video tapes only lasted 20 minutes so they were always switching between tape machines. The night before the cutover, the RCA techs started making the tape switches 1-second early. By the time of the cutover they were running 30-seconds early. The Showtime supervisor didn't realize that anything was amiss until close to 12:30. "Aren't we a little early" he kept asking. The RCA folks assured him that they were right on time and at 12:29:30 the Showtime movie ended and the pictures of Vernon Valley flashed on the screen with the names of all the RCA people rolling down the screen complete with music accompaniment. Every TV monitor in the earth station was tuned to Showtime and when the RCA credits rolled we all cheered. Almost immediately every incoming telephone line started ringing off the hook with Showtime brass wanting to know what the hell was going on -- we cheered even louder.

The early days of the domestic US satellite business were filled with tales of people with big ideas and no money who through skill, cunning and a bit of luck became a huge success. No one epitomized that story better than Bill Rasmussen and ESPN.   Rasmussen had been working at Connecticut television stations when in 1974 he joined hockey's New England Whalers team as Communications Director.   At the end of the 1977-78 season, when the Whalers didn't make the playoffs, Rasmussen and his son Scott, who was the Whalers public address announcer, were both fired.   Casting about for a new venture Bill had an idea about how he could use the new communication satellites to distribute sports programming to cable TV systems.   He incorporated what would ultimately become Entertainment and Sports Programming Network (ESPN)[4] on July 14, 1978.   He began talking to Connecticut cable operators about providing them TV coverage of University of Connecticut basketball and football games and Whalers hockey.   Based on a positive response from the cable guys, he talked to RCA Americom about buying time on F-1 by the hour as needed for game coverage. At that time Americom was scrambling to try to fill up F-1 and get the revenue stream going. They had six F-1 transponders available and were eager to get them sold. Al Parinello, the Americom sales guy pointed out to Rasmussen that he could give him a 24 hour per day preemptible transponder for $35,000 a month that would cover the whole US.   "Anywhere in the U.S.?" asked Rasmussen who had just been thinking about serving Connecticut. "Anywhere" said Parinello.   That not only worked out to be a better price than buying the 5-hours he needed for games at Americom's hourly rates but he would now have national coverage.   The legend goes that Rasmussen paid for the first month with his American Express card and had 90 days to come up with the rest of the annual charge. That was also the last transponder that Americom sold on those terms.   A few weeks later, the Wall Street Journal

---

4 Rasmussen originally called it ESP Network for short but changed it to ESPN when his letterhead came back from the printer that way.

ran a story about how satellite transmission was the future of cable television and all the major players now wanted space on F-1.

Now Rasmussen had a lot of satellite time and had to figure out a way to fill the transponder with programming 24-hours a day, 365 days of the year. According to legend, Bill and his son Scott got stalled in a traffic jam on Interstate 84 on the way to a family birthday party at the New Jersey shore. They started to discuss how they could organize an all sports station with 24-hour programming. By the time they got back home that night they had it pretty well figured out. They also agreed on a half-hour sports news show (now Sports Center) at 6:30 PM opposite, and competing with, the national news shows on ABC, CBS and NBC. They were convinced that there was a large audience intensely interested in sports -- they were right.

The big break came when Rasmussen landed Anheuser-Busch as ESPN's first major sponsor, agreeing to a contract worth $1,380,000. Valentine's Day 1979 was a another big day for ESPN -- Rasmussen signed up the NCAA for ESPN to cover their collegiate sporting events. On the same day he also got an agreement with Getty Oil to buy 85% of ESPN. Getty hired Chet Simmons, president of NBC Sports to be president of ESPN and Scotty Conal, NBC VP of sports operations to be Sr. VP of Operations and Production. Now ESPN was in business. "We will do every single NCAA championship event that the networks don't take," said Rasmussen. "The final four of lacrosse to lacrosse people is every bit as fascinating as the final four in basketball," he said. "The frozen four in hockey, the final four in soccer, we said we are going to do them all. We didn't really know what we were committing to, but we said we would do them" recalled Rasmussen. And they did. They also did Tuesday Night Football from the Canadian Football League, hurling from Ireland, Australian rules football, karate, slo-pitch softball and ping pong. If they ran out of live events they would rerun earlier ones. They bought an acre of land in Bristol, CT for the headquarters that ultimately expanded to a 165 acre campus and started construction on production facilities and a satellite earth station.

I had heard about ESPN but understood that they were under financed, in trouble and didn't pay their bills. I was surprised when trucks from AT&T turned up at Vernon Valley to install a microwave link to connect one of our uplinks to the ESPN facilities in Bristol, CT. It turned out that ESPN had just received the Getty investment and wanted to get on the air in a hurry. Their facilities in Bristol weren't completed when they went on air on September 7, 1979. The first broadcast originated from a TV production truck parked in the muddy construction site. The feed went from the truck through their new AT&T microwave link to Vernon Valley where we put them up on F-1 and the programming went out to 625 cable systems representing about one million viewers. And the rest is history. By 2002 the company's flagship network, ESPN, reached more than 87 million households and televised all of the major professional leagues: baseball, football, hockey, and basketball. According to the 2002 annual report of ESPN's parent company, Walt Disney,[5] ESPN was the number one basic cable network in terms of affiliate, national, and local advertising revenue. Considered by many to be the most successful basic cable network, ESPN delivered the hard-to-capture audience of young males to a wide range of advertisers. Cable system operators consistently selected ESPN as the number one cable network in perceived value.

Sometimes the Americom engineers had to come up with creative solutions to get new customers on the air in a timely manner. Spanish International Network (SIN) started out as a single UHF television station, KCOR Channel 41 in San Antonio, TX. They primarily broadcast Spanish language programming from the Mexican TV network Televisa. The Televisa signal was carried from Mexico City to San Antonio on a Mexican microwave system. Subsequently SIN acquired KMEX Channel 34 in Los Angeles and, in 1968, WXTV Channel 41 in Paterson, NJ. These stations received programming from San Antonio by US terrestrial microwave. SIN very much wanted to be up on an RCA satellite and available to cable systems all over the US as soon as possible. That shouldn't be hard -- after all they were in the same state and only about 27 miles away from the RCA uplink station! An initial survey by

---

5 Getty was the original investor and majority (85%) shareholder in ESPN. In 1984, Getty's interest was bought by ABC. In 1996 Disney bought ABC including ESPN. Hearst remains a minority shareholder with 20% of ESPN.

RCA engineers showed that about two blocks from SIN's WXTV station in Paterson was St. Joseph Hospital, a 12 floor building. It turned out that here was a clear path from the roof of the hospital to the RCA microwave tower on Hamburg Mountain in Vernon. But it would take weeks if not months to procure equipment and install a standard microwave radio link and SIN was in a hurry. The engineers came up with some Farinon portable microwave units that consisted of a three foot dish antenna on a hardwood tripod with the electronics mounted on the back of the dish. These portable units were designed to transmit remote programming, such as sports events from the production truck, back to the TV studio. Remember, this is years before the first satellite news gathering truck. For the WXTV and Hamburg Mountain sites, the tripod was discarded and the dish and electronics were mounted on the tower. The problem was how to install the two units on the hospital roof. One receive unit had to point to WXTV two blocks away and a transmit unit had to be pointed to Hamburg Mountain. The hospital would not allow any fasteners to penetrate the roof. The solution was to put sheets of 3/4" plywood on the roof and bolt the legs of the tripod into the plywood. The engineers were concerned that a strong wind might blow the antenna and the plywood over so plastic garbage cans filled with water were placed on the plywood to anchor it. The receive unit was alined to pick up the signal transmitted from WXTV. The output of the receive unit was cabled over to the transmit unit that was alined on the RCA Hamburg Mountain tower. The signal from Hamburg Mountain then went into the RCA microwave radio link to the Vernon Valley earth station where it was transmitted up to F-1. The system worked like a champ and SIN was now going out to cable systems all over the US. This haywire assembly of portable TV equipment was planned to be a temporary system until a permanent microwave link could be established. As with many such "temporary" arrangements they last much longer than expected. Fall turned into winter and it gets very cold in Paterson. The water in the plastic garbage cans froze and cracked the plastic. When it got warmer the ice melted, the water ran out and the wind blew over the microwave dishes. Very embarrassing! Techs got up on the roof, restored service and weighted down the plywood with sandbags. Work also started in earnest to get the SIN permanent microwave link installed.

In those early days we also had a most unlikely customer on F-1 -- IBM. In the early 1970's MCI, Lockheed and Comsat formed a company called CML Communications to provide digital satellite communications for businesses. In 1974, MCI in need of cash, sold its share of the company to IBM. IBM and Comsat then brought in Aetna Insurance as a partner and renamed the company Satellite Business Systems (SBS). IBM was tasked to design the SBS Time Division Multiple Access (TDMA) digital communications system. IBM needed a satellite transponder to test out the new system so they leased Transponder 23 on F-1. We would look at the channel occasionally on the spectrum analyzer and could see that it was some kind of a digital carrier. One day I got a call from an engineer at IBM at a location near Poughkeepsie, NY. He asked if we were having any trouble with the satellite because they were having intermittent problems with their digital carrier losing lock. I assured him that all was well. He called back a day or so later and said "It just happened again." I again assured him that F-1 was just fine. It turned out that they had set up their earth station antenna in the back of the IBM parking lot right next to the Metro North railroad tracks. Because F-1 was pretty far west out over the Pacific Ocean, at 135 degrees West Longitude, their antenna elevation was low and pointing right across the tracks. Every time a commuter rain came by IBM's transmitted signal was interrupted and reflected back off the speeding train thus causing the digital communications to lose lock

Some time later, I started getting calls from IBM wanting know which way F-1 was moving. I told them I had no idea and would let them talk to one of the orbital analysts. Evidently the rate at which the range to the satellite was changing was screwing up the time synchronization of their digital data. After awhile the analysts started complaining to me about all of the "which way is F-1 going calls" from IBM. Finally, I agreed to give IBM F-1 ephemeris data so they could figure it out for themselves. At one point I had suggested to them that maybe they had invented a system that wouldn't work in a satellite. They ultimately got their Time Division Multiple Access (TDMA) digital system working and launched three Hughes HS-376 Ku band birds. The SBS system, however, turned out to be too expensive for most companies to afford and MCI and Sprint were supplying similar services at lower costs. Comsat left SBS in 1984 and a year later SBS was sold to MCI. The satellite fleet (of by then six spacecraft) was sold off to various operators. I always thought that SBS was another one of those ventures that is started on the basis of some neat new technology without a lot of thought about being given to how it might make money in the real world.

As succesful as Americom was, didn't mean that there weren't some bumps in the road. When RCA signed up Showtime for two transponders and agreed to build a production facility for them at Vernon Valley, HBO complained about the terms of the Showtime deal. HBO charged that the Showtime transponders cost Viacom less than what HBO paid for their two transponders. HBO paid the full tariffed rate for transponders 20 and 24, however, they only used them for 12 hours a day. Showtime on the other hand only contracted for and paid for the actual 7-hours a day that they used transponders 4 and 10. HBO was sufficiently miffed that they started discussions with Western Union about leasing two transponders on Westar 2. HBO and Showtime had both contracted for "protected" service which meant that RCA guaranteed that they would restore their service on another transponder or another spacecraft in case of a failure. The cable magazine CATJ reported that Western Union offered HBO two transponders at the FCC tariffed rate for "preemptible" service (much cheaper than "protected") with a wink and a nod that HBO would never be preempted. In fact Western Union also promised that they would launch Westar 3 to assure that they had sufficient backup transponders in case of failures. HBO of course didn't want be alone on Westar 2 and leave Showtime, WTBS and CBN on the RCA bird. Cable operators only had one antenna and they sure weren't going to buy another one just to keep HBO when they had access to Showtime on F-1. Western Union was trying to put together a deal to get HBO, WTBS and CBN to come over to Westar-2 thus making it the defacto "cable bird" and leaving Showtime on F-1. One of the problems with that deal was that the RCA contracts contained penalty clauses that would cost HBO $534,000, WTBS $462,000 and CBN $722,000 if they left RCA before their contracts were up. How those penalties would get paid was certainly a factor in the negotiations. HBO's president Jerry Levin characterized the negotiations as a "business decision for the whole industry" thus encouraging the perception that "as HBO goes, so goes CATV". Levin further stated that HBO was "in the marketplace looking for a 10-year contract with some satellite operator". Cable magazine CATJ labeled Levin as the "father-negotiator" when he stated "We are looking for the kind of arrangement which will enable the industry to take advantage of technological changes which may come about." Showtime was notably silent through all of this.

Andy Inglis certainly wasn't taking all of this lying down. In the late fall of 1977, Andy had used the RCA helicopter to bring he and Levin and a couple of other HBO executives up to Vernon Valley from the HBO offices in Manhattan. After I had given the usual $10 dog and pony show to the HBO entourage they retired to our conference room for about an hour and a half. As they were departing for the heli-pad, Andy's comments to me sounded like HBO was leaning toward Westar -- not good. Andy set up another meeting with HBO executives including Gerry Levin and Ed Horowitz for January 18, 1978 at the Americom offices in Princeton. As the meeting began, it started to snow and by lunchtime it was snowing heavily. Andy seizing the opportunity, called his wife Marie and told her that we are going to have overnight guests tonight. Andy was driven to work everyday from his home in nearby Moorestown in an RCA Lincoln Town Car. He called in his chauffer and said "call Marie and find out what food and drink she needs, go out and get them, and deliver them to my house". As the afternoon wore on, discussions were stalled. Andy convinced the HBO contingent that it was too snowy to return to New York and that they could all stay at his house that night. Andy Inglis was a true gentleman and I am sure the perfect host. In any event the snowbound group continued the discussions late into the night and the next morning Andy announced to close associates that he had a deal. On January 30, 1978 HBO and RCA announced that they had come to terms for the on-going future use of the Satcom spacecraft. Also included in the deal was a contract for a third transponder for what would ultimately would become Cinemax. A collective sigh of relief was heard throughout Americom. As a result of these negotiations

RCA also established the "grow with" tariff that gave new services a reduced rate of as much as 50% in the first year and then is increased over the years as the business grows. The total revenue under the 10-year contract was about the same as the old fixed price tariff but the lower "going in" rates provide more new business opportunities.

An interesting footnote to all of this was Western Union's offer to HBO to launch Westar-3, their ground spare, as part of the deal to lure HBO away from RCA. In 1976 Western Union (WU) had entered into a contract with NASA to build and operate NASA's geosynchronous Tracking and Data Relay Satellite System (TDRSS). TDRSS prime mission was to relay data from NASA low orbit satellites thus allowing NASA to close expensive overseas tracking stations. Also included on the TDRSS spacecraft were 12 C-Band transponders (Advanced Westar) that WU planned to use for their commercial customers. The decision to change TDRSS launches from the expendable Atlas launch vehicle to the Space Shuttle, however, put the TDRSS program way behind schedule. TDRS-A was supposed to launch in 1978, was postponed a number of times and finally launched in 1983. In 1978 WU sued NASA because their C-Band transponders would not be available when required by the contract. WU won the suit and NASA effectively gave Westar-3 a free Delta launch on August 10, 1979 -- no wonder Western Union offered cheap Westar-3 transponders to HBO so easily!

By December 1978 RCA Americom's business was doing very well. Americom President Andy Inglis announced plans to launch a 3rd satellite in December 1979, about a year ahead of schedule. He also announced that "We are rapidly approaching the break-even point and we expect to be in the black early in the first quarter of 1979". Andy said that revenue for 1978 would be a 52 percent increase over 1977's and that next year he expected a 47 percent increase. The National Cable Television Association said that 5 million homes were connected to cable systems that were connected to 900 satellite earth stations. Based on these growth figures Americom planned to launch a fourth satellite on the Space Shuttle in the early '80's. In three years we had come a long way baby.

Chapter 10 - Big Troubles In River City

Any time a new technical system goes into use, unforseen problems often occur in its operation. When a product is as innovative and complex as the Satcom spacecraft and is operated in the hostile environment of outer space for the first time, such problems are almost inevitable. Also, a spacecraft designed to be as light as possible and live in the rigors of outer space also has to be able to survive the severe vibration, pressure and acceleration of launch. All of these factors pose severe design challenges so it is not surprising that there may be some problems with the spacecraft once it is launched and on station. With the Satcoms we certainly had more than our share of spacecraft anomalies. Some were caused by design miscalculations or performance factors that could not be tested in the earthbound 1G environment. Others were caused by manufacturing problems either at Astro or their many subcontractors. No matter what problems we had or how vexing they were, however, these spacecraft exceeded their designed lifetime, provided our customers outstanding performance and made RCA a lot of money. Being fairly subjective about all of this, I believe that a lot of the credit for that goes to Spacecraft Operations and how we worked around these problems.

In September of 1977, we had a serious problem. My home phone rang in the middle of the night and the on duty Spacecraft Controller told me "F-2 has lost lock on the earth." I told him "Send command list Pitch Off -- have you called the Duty Analyst?" He told me both analysts were on their way in. I pulled on my clothes, jumped in my car and started down to the station. "Down" was the operative word. I lived in Highland Lakes which is about 600 feet higher than Vernon where the Vernon Valley Earth Station was located. The most direct route was on, appropriately named, Breakneck Road -- it was a winding 25% grade for a mile. It was really exciting in the winter. When I arrived in the station, F-2 was locked up on the earth and operating normally. It had evidently come back when the Pitch Off command list was sent. We and the engineers in both our Princeton headquarters and Astro pored over the telemetry data before during and after the event and could find nothing out of the ordinary. The consensus seemed to be that it was some kind of a random event -- and then it happened again.

The Pitch Off command list was sent again and the spacecraft recovered. Of course every time this happened, we lost communications and the phones lit up from our customers. After the second event our customers were getting a little nervous -- so was I. The general consensus now was that it was an intermittent problem in one of the components in the Pitch Control Loop -- but which one? The easiest unit to switch was the Earth Sensor Assembly (ESA). We switched ESA's and a few days later again lost pitch lock. We replaced the Attitude Control Electronics 1 (ACE1) with ACE2 and shortly thereafter we lost pitch lock again and were able to recover as before. The next logical choice of possible culprits was the Momentum Wheel, however, switching Momentum Wheel Assemblies (MWA) is difficult. The spacecraft loses lock on the earth in the process of the switch but it appeared that we had no other choice. I stood directly behind the Vernon Valley Spacecraft Controller as we started this evolution. The first command list set the replacement MWA2 to 200 RPM warmup speed. The online wheel MWA1 changed speed to compensate for the additional momentum from MWA2. When MWA2 was warmed up, we commanded it to ON which spun it up to 6000 RPM. In the same command list we commanded MWA1 to 200 RPM which braked it and slowed it down. At about this point, we lost pitch lock and communications were interrupted. We had warned our customers what we were doing and to expect a short outage. As the momentum was exchanged between the two wheels the spacecraft started to turn back toward the earth. As MWA1 got down to 200 RPM, we commanded it off. The spacecraft came around and locked up on the earth. We were now operating on MWA2. I remained behind the controller observing the data and making sure that everything was OK. There was a small amount of roll in the bird so I told the controller to command the Magnetic Roll Controller ON. I was watching the commands go out. The instant that the Mag Roll Controller 1 ON command went out the Momentum Wheel Speed dropped and we lost Pitch Lock. "It's the Mag Roll Controller" I yelled "command Magnetics OFF". Shortly after the magnetics went off we regained pitch lock. We immediately put out an Operations Bulletin for F-2 stating that under no circumstances was Magnetic Roll Control 1 to be used until we got all of this sorted out. We very carefully verified that Mag Roll Controller 2 worked

124

normally.   We had no more Pitch Lock loss incidents.

It was very puzzling because the Magnetic Roll Controller is not part of the Pitch Control Loop and shouldn't affect the MWA.   It turned out however, that a 10 volt Power Supply in the Magnetic Roll Controller, in addition to powering the magnetics, also provided a critical voltage for the MWA.   Something in Magnetic Roll Controller 1 was shorted circuited and when it was commanded on it shorted out the 10 volt power supply causing the MWA to immediately spin down.   The wheel's momentum was then transferred to the spacecraft body which caused the spacecraft to start slowly spinning and lose lock on the earth. This was very troublesome because we used magnetics to control roll all the time and now we no longer had redundancy.   Our troubles turned out to be far worse than that.

 The Earth Sensor Assembly (ESA) provides Roll Error data and Pitch Error data for spacecraft attitude control.   It cannot, however, provide Yaw Error data.   In normal operations this is not a problem because every 6-hours Roll and Yaw interchange.   If there is a Yaw Error, six hours later it appears as a Roll Error and can be corrected.   During North/South maneuvers, however, there can be large yaw errors that have to be corrected immediately.   For that reason, the spacecraft has a Yaw Gyro that is only turned on for North/South maneuvers. Unfortunately, the Yaw Error data from the Yaw Gyro goes through the Magnetic Roll Controller.   We had already lost Mag Roll Controller 1 because of the shorted 10-volt Power Supply.   If we lost Mag Roll Controller 2, we would lose the capability to do North South Maneuvers -- and that would be catastrophic.   Spacecraft Engineering started working on contingency procedures to do N/S maneuvers manually in case the other controller failed.   Joe Elko, Americom's Spacecraft Engineering Manager, came up with a control algorithm to manually fire north face thrusters in a sequence that drove the spacecraft in the desired direction while correcting roll and yaw errors induced by the firing.   We built the command lists to implement Joe's firing algorithm and tested them on F-2.   We also built command lists to do roll and yaw controls with the north face thrusters.   It was fortuitous that Spacecraft

Engineering did an excellent job in devising a solution and that we made the required preparations because Mag Roll Controller 2 ultimately failed.

In anticipation of a possible failure, we had tightened up F2's North South limits. Instead of letting the spacecraft go out to a 0.08 degree latitude error we would do a North/South at about 0.06 degrees and drive it out to about 0.09 on the "high" side.. That way, in the event that the Mag Roll Controller failed, we would have a longer time to test various procedures before the bird was out of the box. When the other Roll Controller failed, we had our procedure in place and the necessary command lists built and now we had to find out if all of this was going to work. Joe Elko came up to Vernon when we tried the manual North/South for the first time. The first thing we had to do was position the Solar Array in what we thought was the right position based on history. There was no sense doing a plume test because our manual N/S was just like a plume test -- we would find out pretty quickly if we were in the right place. We also found out pretty quickly that array position was much more critical for this procedure than for a regular North/South. We had a strip chart recorder (SCR) set up displaying Roll (on two scales), Pitch and Compensation Amp voltage which showed when we needed a Momentum Adjust. Joe Elko's North/South maneuver command list fired north face thrusters 1 and 4 for 3 seconds, thrusters 1, 2, 3 & 4 for 11 seconds and then 1 and 4 for 3 seconds. The list waited for 90-seconds (half nutation period) and fired the same thruster sequence again.

Hopefully the spacecraft wasn't too disturbed and we could repeat the north face pulse pair again. More often than not, however, the spacecraft was disturbed so we would observe the roll on the SCR and decide what to do next. Maybe the solar array needed to be moved to a better position. Maybe we needed to correct roll. Maybe we needed to kill nutation (bobbing up and down). It was all very subjective and depended on the skill of the analyst directing the maneuver. Not being able to use the yaw gyro, of course, we couldn't see how much yaw the firing had put into the spacecraft. Bob Youngblood devised a plastic template that we could lay down on the SCR Roll trace and measure Roll Rate giving us an indication of Yaw. He had marked the template with lines indicating the equivalent degrees of Yaw for each Roll Rate. We would use various north face thruster combinations to correct roll, remove nutation or remove yaw. Because we were using North Face thrusters for everything, we got some North/South correction from everything we did including correcting disturbances. Even so, it was impossible to get sufficient N/S burn at a single node crossing. It meant that the process had to be repeated over 5 to 7 days.

Because this procedure put fairly big disturbances in the spacecraft, I wanted to be sure that we were not degrading customer traffic. Miami was on the edge of F-2s communication "footprint" and thus more sensitive to disturbances. I set up a Frequency Selective Voltmeter tuned to the Miami PLC carrier to measure Idle Channel Noise (ICN). As long as the ICN was within spec, we were all right. I was pretty amazed at how forgiving the spacecraft was -- sometimes we had disturbances of one-degree without serious degradation of traffic.

Doing North/Souths this way was not easy. In real time the analyst had to analyze data and make immediate decisions about firing thrusters. A wrong judgement could result in a severe disturbance and interruption to all the F-2 traffic (including Alaska). It was a grueling, tedious, time consuming procedure but it worked and enabled F-2 to operate effectively and produce revenue for five years after the Roll Controller failure. Unfortunately Americom management didn't appreciate the dedication and skill required of the analysts. I submitted the names of the analysts that did the F-2 North/Souths for a Technical Achievement Award. The Vice President Technical Operations replied that they were "just doing their jobs".

Late in life, a command bit line failed in F-1 making it impossible to command the array to stop at the dawn or dusk position over the thrusters for North South maneuvers. The only way a North South maneuver could be accomplished was to fire thrusters while the array was moving. Analysis of several years of plume test and N/S maneuver data yielded the earliest time prior to ascending node that a North/South could be started without severe attitude disturbances. The procedure was to start the maneuver then and let the onboard logic abort automatically when yaw errors exceeded limits. Using that procedure we could get about 10 to 12 minutes burn a day. Because the North/South maneuver time only coincided with the array dawn or dusk position twice a year, we were doing North/South's at the wrong time of day most of the year. A great deal of North/South trajectory planning by the analysts was required to keep F-1 in the box without wasting fuel. This was another tedious, time consuming procedure but it worked and kept F-1 bringing in revenue -- and we were just doing our job.

As noted earlier, the REWIND procedure had damaged the F-2 solar array causing loss of power. As F-2 approached the end of life, a number of transponders were turned off because of insufficient electric power. We also turned off many heaters so that we could keep the maximum number of transponders on. One day, the spacecraft controller was executing a routine East/West maneuver and he called the analyst and reported that East West Start command list had executed normally but nothing happened. Under an analysts direction, East/West Start was executed again -- sure enough nothing happened and the thruster temperatures didn't rise. Checking various temperature sensors showed that the spacecraft was quite cold -- we had frozen the fuel lines! We turned on one of the transponder heaters near the fuel lines and the next day successfully executed the East/West. We were lucky -- the fuel lines were frozen to all four thrusters we were trying to fire. If one or two of them had fired we would have had a severe attitude disturbance. Clean living pays off again!

Chapter 11 - A Flashback -- American Satellite Corporation

I have mentioned that I worked for American Satellite Corporation for a year in 1974 and 1975. In many ways that year was very difficult, however, I also had an opportunity to learn a lot about satellite communications.   So I will interrupt the Americom narrative to flashback to the Spring of 1974 when I was working for RCA Service Co. at NASA's Goddard Space Flight Center (GSFC).   I was Manager, Technical Support, one of three managers in the RCA STADAN Operations Support Group.   STADAN was the acronym for the Space Tracking and Data Acquisition Network, NASA'a worldwide network of fifteen Satellite Tracking Stations supporting Scientific Satellite Programs.   For the previous eight years I had great job.   I was responsible for the Network Support Team (NST), a 24/7 team that kept track of everything that affected spacecraft operations in the 15 station network including problems with the 50 satellites that the stations were supporting.   I also had two engineering groups; one that designed and implemented modifications to station hardware and software, and another that provided general engineering support for NASA's Network Operations Branch.   I also had a group of PhD level engineers that supported NASA's Operations Evaluation Branch.   They worked on projects to determine errors in the various satellite tracking systems (Minitrack, Optical, Range & Range Rate and Laser) and how to correct them.   It was a good job and I looked forward to going to work each day.   My problem was that RCA Service Co. had just lost the GSFC Operations Support contract to Bendix Field Engineering and I had laid off about 50 people.   I still had a job and the Service Co. was assuring me that they would find a place for me, however, I was pretty fed up with ups and downs of government service contracts.

One Sunday I saw an ad in the Washington Post for a Network Operations Manager for the American Satellite Corporation.   Now, although I knew a lot about scientific satellite operations, I didn't know very much about commercial communications satellites.   I hiked across the Goddard campus to the huge GSFC Library.   I was in luck -- the American Institute

129

of Aeronautics and Astronautics (AIAA) had just finished up their 5th Communications Satellite Systems Conference at Los Angeles in April and the GSFC Library had all of the conference papers. And better than that, almost all of the management of the newly formed American Satellite Corporation (AMSAT) had presented papers. I read and digested them all and went home and updated my resume to emphasize how my qualifications would fit with AMSAT's mission. I got the job.

AMSAT (also known as ASC) was a small startup company and its business plan was based on selling Private Leased Channel (PLC) telephone service and digital data transmission. The Vice President of Engineering and Operations was Dr. Eugene Cacciamani a world renowned digital communications expert. The AMSAT technical staff was convinced that although their PLC system was completely analog now, digital was the only way to go as soon as the technology advanced and the costs came down. ASC had been formed by Emanuel Fthenakis in 1972 with Fairchild Industries as the major stockholder. On March 20 1973, ASC announced a $25 million contract with Hughes Space and Communications Co. for three 12-channel HS-333 communications satellites. ASC also gave a down payment to NASA for a first launch on the Space Shuttle in the 4th quarter of 1974. By then they planned to a have a network of eight earth stations serving New York, Dallas, Chicago, Washington, Atlanta or Miami, Los Angeles, San Francisco and Seattle. They had also initiated a project with Fairchild Space and Electronics for design and development of an advanced 24 transponder spacecraft. By the time that I arrived at ASC in June 1974, their plans had been scaled down considerably. Instead of building and launching their own satellites, ASC had leased two transponders in Western Union's Westar-2. The Spacecraft Engineering group was winding down operations and and filing away all their data for possible future use. They had only three earth stations serving New York, Dallas and Los Angeles.

Manny Fthenakis had come from Comsat and ASC was heavily staffed with Comsat and Intelsat alumni -- at that time about the only people in the world with communications satellite

130

operations experience.   Comsat was the United States partner in Intelsat and operated all of the Intelsat earth stations in the US.   Anyone in ASC not having a Comsat/Intelsat heritage was looked on with some suspicion if not disdain -- which I discovered rather quickly.   Despite the fact that I had spent the last nine years working for RCA at Goddard Space Flight Center, I was considered an outsider by this closely knit group of satellite communications folks.   My first two weeks with the company was spent at ASC headquarters at the Fairchild Space plant in Germantown, MD for indoctrination. My semi-permanent location would be at the ASC earth station in Vernon, NJ.   The plan was that after a year at Vernon, I would return to Germantown where a Network Operations Center would be established.   During my two weeks at Germantown, I quickly realized that there were a number of power struggles going on is this newly organized company.   Although most of the people were from Comsat, there was a definite division between Comsat Headquarters types and the Comsat Field Operations troops. Unfortunately I had been hired by the headquarters group and was being sent to an earth station peopled by field folks.   On top of that the Vernon Earth Station Manager, Wil Zarecor, did not see the need for a Network Operations Manager and in fact thought that he should have my responsibilities. Even after I was on board in Germantown, Wil continued to object to the arrangement as I was to find out as I journeyed to Vernon.

I drove from Lanham, MD to Vernon on a Sunday afternoon in late June and checked into a motel. The next morning as I was eating breakfast, the motel manager relayed a message to me that said "Don't go to the Vernon Earth Station until I let you know". It was signed George, my boss. About 10:30 George called and said that everything was OK and that I should go to Vernon.   I entered the earth station with a fair amount of trepidation and introduced myself to Wil Zarecor.   Wil was clearly not happy about all this and told me so, however, he was a gentleman and we both agreed that we would make the best of things.   Will introduced me to Jim Evans the station engineer who was also not particularly happy about my being there. Wil, who had been at the Brewster Flats, WA Intelsat station, was a very smart guy and an excellent manager. After some initial rough spots we became good friends. Evans, who had

131

been at the Andover, ME Intelsat station, took a while longer.

I had a desk on the operations room floor right in the middle of all the communications equipment. Besides the Vernon station, which served New York, ASC had an earth station at Nuevo, CA serving Los Angeles and Murphy, TX serving Dallas. By this time, although they were not yet carrying traffic, all three earth stations had established message carriers in Westar-2. There was an order wire system in each carrier that allowed telephone and teletype communication between the three stations. There was also a land line connection extending the order wire to headquarters in Germantown. One of the first responsibilities everyone agreed that I should have was the order wire. I soon found out why -- it didn't work worth a damn. It turned out that PulseComm order wire equipment was very sensitive to signal level changes. If the test tone levels were not exactly +7 and -16 dB it screwed up the Echo Suppressors. On the voice circuit echos were just annoying but echos drove the teletype equipment nuts -- all the messages would be garbled. ASC was very dependent on the teletype for hard copy messages to the earth stations so there was a lot of pressure to keep the order wire working. It seemed that about once a week I had to oversee an end-to-end alinement of the order wire between all three stations and Germantown.

American Satellite had advertised that they would be "up and running in July". These ads had appeared in many of the trade publications and also in the Wall Street Journal. The sales people had sold some PLC circuits and now the technical people were working to get the them up and operating. During July, I had been busy scheduling and supervising end-to-end tests of the FDM Supergroups and telephone carriers between the Dallas, New York and Los Angeles Central Offices. They had gone well but by the last week in July we still didn't have any circuits connected to customers at both ends.   The most promising were a couple of New York to Dallas circuits for Smith Barney, the Wall Street stock brokers.   The local telephone loops were in at both ends and the customer packages had been installed and alined but there were still problems with the circuits.   Technicians had worked over the weekend of July

27/28 and still none of the Smith Barney circuits were working. AMSAT management was frantic because their promise to be up in July was a big deal and the company would be seriously embarrassed if we didn't make it. By the afternoon of Wednesday July 31, one of the Smith Barney circuits was up but it was noisy. Work continued into the evening. I wasn't going home until I knew the outcome. It kept getting later and there still was a problem at the interface of the AMSAT and Southern Pacific (SP) microwave in Orange County, NY. Finally at about 11PM, with only an hour to go, the circuit came good. The AMSAT techs in Smith Barney in New York were talking to the AMSAT techs in Smith Barney in Dallas -- it was working. Not so fast the sales guys said -- this doesn't count unless we have the customer use the phone. At that point all the Smith Barney people had long ago gone home and we literally had people holding wires together to keep the circuit working. AMSAT techs at both ends ran through their buildings looking for anyone that could be put on the phone. Eventually we ended up with a couple of cleaning ladies from the night crew -- one in New York and one in Dallas talking to one another at about 10 minutes to midnight July 31, 1974. Up and running in July indeed! In the next few days the circuit problems were permanently resolved and two circuits from New York to Dallas for Smith Barney were the beginning of AMSAT's PLC business.

One of the reasons for all of the concerns about AMSAT being up and running in July was because AMSAT had a pending proposal to Dow-Jones to transmit the Wall Street Journal to their regional printing plants by satellite. Part of the deal was a full page ad, extolling AMSAT's satellite business, to be run in the Wall Street Journal, the Los Angeles Times and the Dallas Morning News -- AMSAT's three markets. The second line in the ad in bold type said "As promised American Satellite is up and running." The headline on the ad in very large bold type said "This ad is printed from a satellite signal". During the last week in July, Dow-Jones, who owned the Journal, had shipped professional grade facsimile equipment weighing about 1000 pounds to AMSAT's Nuevo, CA earth station and to Vernon. This fax equipment could transmit a high resolution image of a 15" by 23" newspaper page. The fax machine included equipment to process and output a full size photographic negative that could be used to burn an offset plate to print the newspaper. The plan was to have Nuevo transmit the AMSAT ad page to Vernon where the negative would be produced. We were running out of time. Full page space had been bought in the three newspapers for Thursday August 8, 1974. All of the Dow Jones equipment had been delivered and checked out by Friday evening August 2nd.

The Dow-Jones people arrived bright and early on Saturday morning in Vernon and our break room was converted into a darkroom to load the film. Everyone was pretty nervous because this had to work. Wil Zarecor and Jim Evans had taken charge of the Vernon operation and the station manager and engineer in Nuevo (both former Intelsat earth station guys) were handling things there. My role at this point was observer. Nuevo sent the first test page and a paper proof was printed out in Vernon. Wow -- there it was the AMSAT ad just as it would look on the news page -- so we thought. One of the Dow-Jones guys laid the page out on the table and brought over his large metal 90-degree square. He alined the square with the right edge of the image and slid the square down until the metal edge was just touching the top edge. The top edge was not square -- the image was skewed! The Dow-Jones people were adamant -- their name wasn't going on it unless it was perfect. Several more tests were run including one producing a negative -- it still wasn't square. Finally, the Dow-Jones fax guy said "I think that the problem is that the audio oscillators in the Nuevo and Vernon fax units aren't the same frequency. The amount of skew is a function of the difference in frequencies." The station mangers and engineers conferenced on the order wire to try to figure out what to do. Neither station had a precision counter that could measure audio frequencies. They tried tweaking the oscillator frequency and making another run to see if the skew became less. It only seemed to get worse. By this time it was about two PM and it seemed to me that they had run out of ideas.

This Full Page Is Printed From a Satellite Signal

As promised American Satellite is Up and Running

The words and pictures you now see on this page were transmitted by facsimile communications signals. The signals were beamed from our Los Angeles earth station via satellite—to our New York earth station using the facsimile equipment of Dow-Jones, Inc. The complete transmission—a distance of 44,600 miles—was accomplished in just 4 minutes! **In fact, this entire page is actually printed from the image received in New York.** It has not been retouched in any way!

American Satellite can produce these same results for all forms of private line communications. Our system has the full capabilities to transmit voice; high or low speed data, and facsimile signals—to and from business and government centers across the country. And with significant cost savings as compared with conventional communications methods.

Right now we're transmitting between New York, Los Angeles, and Dallas. Chicago and San Francisco will be operational later this year. Other points nationwide will be added soon.

Reliable, low-cost domestic satellite communications—the new pacesetter for American business—is already introducing revolutionary changes you simply can't afford to have pass you by. American Satellite is now operational and ready to meet your most particular communications *and* budget requirements. Contact American Satellite—the foremost name in the domestic satellite field—today! Call or write Dennis T. Goddard, Vice President, Marketing.

american satellite corporation

20030 Century Boulevard Germantown, Maryland 20767 301/428-0100

This is the ad that was transmitted from Nuevo,CA via AMSAT's transponder on Westar-2 to Vernon, NJ on August 3, 1974 and printed in the Wall Street Journal, Dallas Morning News and the Los Angeles Times on August 8, 1974 - *American Satellite Corp.*

At that point I called Nuevo on the order wire and asked if they could run a cable from the fax oscillator test point to the modulator on their uplink to Vernon -- they said that they could. This way they could send their oscillator signal to Vernon. I asked Wil if we could get a cable from our Nuevo downlink over to the fax machine and connect an oscilloscope to it. We connected our fax oscillator test point to the vertical direct input and the Nuevo signal to the horizontal direct input of the oscilloscope. We then had a classic Lissajous pattern on the scope and Jim adjusted our fax oscillator frequency until the Lissajous pattern became a perfect circle that meant both frequencies were identical. There were a lot of "why didn't I think of that" looks around the ops room. The techs normalled everything up, Nuevo transmitted the page and when the Dow guys checked, it was perfectly square. They ran another one and made a perfect negative that the Dow guys took away for the ad. After this, I didn't exactly become one of the Comsat guys, however, there was certainly a bit more respect for my technical ability and ingenuity.

The ASC Vernon Earth Station was connected to the ASC Central Telephone Office in the World Trade Center Tower 2 in New York by a microwave radio link. Because ASC did not have sufficient time to design their own microwave system and get FCC clearance, they made a deal with Southern Pacific (SP) Communications to connect the New Jersey earth station to the New York central office. SP built a microwave link from the Vernon earth station to a point in Orange County, NY to intercept SP's backbone microwave system running from Syracuse, NY to Jersey City, NJ. At the SP site on Tonnelle Ave in Jersey City SP built a 4GHz link to the roof of Tower 2 of the Trade Center.

ASC's Central Telephone Office (CTO) in New York was located on the 110th floor of the South Tower. The South Tower was still under construction at that time so going to the ASC CTO was quite an experience. The building was completed and occupied up to the 78th floor. After taking the PATH train over from New Jersey, I walked up to the Tower 2 lobby and took the elevator to the 78th floor. It was a fast trip -- the elevators traveled at a rate of two floors per second.   When I arrived at the 78th floor, I had to log in with the construction contractor and a Port Authority supervisor.   The World Trade Center buildings were being built by the Port of New York Authority (PONYA).   I was then allowed to board an elevator that consisted of a plywood box with no door with a union elevator operator.   It went as fast as the other elevator and the floors zoomed past the door opening.   I flattened myself against the plywood back of the car.   I got off the elevator at the 107th floor and gingerly walked down the center of the hall -- there were five more open elevator shafts with only a 2 by 4 nailed across the opening.   I felt that one trip or misstep would result in a 107 floor plunge.   The 107th floor was completely unfinished.   There were no walls except those around the elevator shafts. The stairs to the upper floors were just cast concrete steps with no walls or rails.   When I got to the 110th floor the ASC Central telephone Office was the only finished area on the floor.   It was in the southwest corner of the floor near the escalator going to the roof and the yet-to-be-built observation deck.   It consisted of an operations room, a battery room and the managers office.   The operations room had several racks of microwave equipment, a half dozen tall racks of multiplex equipment and some telephone racks with punch-down blocks for telco local loop termination.   All of the microwave and mux equipment used 24-volt DC power.   This equipment was powered directly from a large battery bank and the batteries were kept charged by redundant battery chargers running off ConEd commercial power.   There was a DC to AC inverter to supply 120VAC power for test equipment and other ancillary equipment in case of a commercial power failure.[6]

Getting the batteries (which are very heavy) from ground level up to the 110th floor was another interesting experience.   None of the passenger elevators went directly from the first to the 110th floor.   As I did, you had to change elevators at the midpoint.   There was, however, a freight elevator that went from the loading dock area to the 110th floor.   The AMSAT technicians were helping to load the batteries into the elevator.   As the heavy batteries were

---

6   One Saturday morning I got a call from the CTO that they had lost power and were running on batteries. The circuit breaker for our facility had tripped and the PONYA people couldn't find which circuit breaker was ours.   We were without power for about 12 hours before PONYA finally found our breaker -- fortunately we had plenty of battery capacity.

loaded, the elevator kept getting lower and lower. First it was an inch below the door sill and and then two. The operator had to keep raising the elevator slightly. Obviously the 110 floors of elevator cable was stretching under the heavy load. When all of the batteries were loaded, the union elevator operator told the techs to get in for the ride up to the 110th floor. To a man, the techs said they thought that they would rather use the passenger elevator.

The first time I went to the CTO, the manager took me up to the roof to show me the microwave antenna. Initially people working there had free access to the roof. Now PONYA had a guard checking ID before we could walk up the yet to be activated escalator. A few months before, on August 7, 1974, the French high wire artist Phillippe Petit had, after months of planning with the help of compatriots and union construction workers, strung a 3/4" wire between the South Tower and the North Tower. He then started from the South Tower and walked across to the North. He danced and jumped and generally amazed everyone with his feats. The NYPD finally convinced him to come in and surrender. The first time that I was on the roof, they were just building the track around the four sides of the roof for the window washing machine that would lower a scaffold for the window washers. We walked over to the edge and looked down -- it made my knees ache. Of course, when the building was completed and the observation deck built there was an electric fence to keep would-be jumpers from getting anywhere near the edge. AMSAT's large microwave horn antenna was located on a 10-foot tower near the west side of the roof pointed toward the SP microwave facility 4-miles away in Jersey City. Because there was so much 4GHz microwave communications in Manhattan, a horn antenna had to be used because its narrower beam reduced the possibility of radio interference. About eight months later when AMSAT built its own microwave link to Vernon, they found out how heavy the microwave usage in Manhattan was. The microwave radio link went by 11GHz microwave from the Vernon earth station 11 miles to a tower on Bellevale Mountain in Orange County, NY and from there the link went 42-miles by 4GHz microwave to the horn antenna on the roof of the World Trade Center South Tower. The plan to cutover from SP involved substituting a small dish antenna for the horn on the SP route to

Jersey City at about 2AM.   Although the dish beam was wider than the horn, Jim Evans figured that the cutover would only take about a half hour so it would be all over before anyone noticed any interference at that hour of the morning.   Jim and the techs switched antennas using lengths of flexible waveguide and then turned the horn around to the predicted azimuth to Bellvale, NY.   Jim had a spectrum analyzer connected to the horn and was looking for a carrier on Bellvale's 4GHz frequency.   He quickly found one and called Bellvale to turn their carrier off -- they did and the carrier didn't disappear -- obviously someone else's radio on that frequency.   He moved the antenna slightly and found another carrier on the right frequency -- Bellvale turned off -- carrier still there!   They moved the horn again and again without finding the Bellvale carrier.   Finally as it was getting light in the East they found the Bellvale carrier, peaked up on it and switched the traffic over to AMSAT's microwave link.   What was supposed to take a half hour took more like four while they sorted out all the 4GHz carriers transmitting into Manhattan.   Fortunately there were no complaints about interference.

When the Port Authority built the elevated observation deck on Tower 2, it turned out that the escalator housing would block AMSAT's horn.   The only solution was to raise the horn by adding another section to the tower.   By this time AMSAT was carrying quite a bit of PLC traffic and couldn't tolerate any long microwave outages.   Riggers built a heavy wood structure around the base of the tower and installed hydraulic jacks to lift the tower and the horn.   This was another 2AM evolution that started with the quick addition of a section of flexible waveguide the same length as the tower section.   The tower was unbolted from the base and the riggers then slowly and evenly jacked up the tower without interrupting traffic. They then slid in the new tower section and gently lowered the tower and horn down on to it.   After everything was bolted down, Jim tweaked the horn position to maximize the Bellvale carrier. There had been a few momentary outages when the waveguide pieces were put in but overall the traffic was maintained.   It seemed like life at AMSAT was just one adventure after another.

There was yet another adventure in the saga of the Bellvale, NY to World Trade Center microwave radio link. That was a long path for C-Band microwave over some fairly mountainous (by Eastern standards) terrain.   The microwave was constructed during the winter and required a helicopter to get the horn antenna to the top of the tower on Bellevale Mountain.   In spring, just as the leaves were coming on the trees, the microwave link started to get noisy.   The noise in the order wire sounded like the surf was up.   We were all sure that the path had not been calculated correctly and that there were trees on some mountain top in the microwave path.   All of the PLC traffic that ASC had was carried in that path -- panic ensued! Did the engineers have the tower heights right?   Was Bellevale Mountain really as high as the geodetic maps said it was?   Wil Zarecor, the AMSAT Station Manager, borrowed a geodetic altimeter from Rutgers University to verify the altitude of Bellvale Mountain at the base of the tower.   To calibrate the altitude function, it had to be placed on a First Order

Survey Geodetic Monument.   It turned out that the nearest First Order survey marker to Bellevale was in the middle of the road at the Upper Greenwood Lake bridge.   It ended up with a couple of AMSAT techs holding up traffic while Wil crouched over in the middle of the road adjusting the altimeter.   Then into the car and a mad dash over country roads to get to Bellevale before the barometric pressure changed -- the altitude appeared to be right. Fortunately, the problem turned out not to be a bad path but some water in the antenna waveguide in New York.

While all these adventures were kind of fun and I was learning a lot, AMSAT was in turmoil and I was getting very nervous.   Every couple of months there would be a reorganization. My boss George was fired and then I was working for the leader of the former spacecraft group. A couple of months later I was working for someone else and then someone else.   I got the feeling that I was being shuffled around because they didn't really want me but didn't know how to gracefully get rid of me.   When I asked questions about my future with ASC I didn't feel that I was getting straight answers. Fortunately for me, in June of 1975 Charlie Twitty at RCA Globcom called me and asked if I would like to come back to RCA -- boy did I ever!

AMSAT was an interesting company. They had a large number of really smart technical people but, in my estimation, suffered from poor management. As their PLC business increased, they were rapidly growing out of their two Westar transponders. A decision was made to make a rather large investment in syllabic compandors for each of the PLC circuits that would reduce the satellite bandwidth required by about threefold. Using compandors they would not need additional satellite capacity for the forseeable future. Then a couple of years later, AMSAT decided to buy a share of the Westar 4 and 5 spacecraft -- now they had all kinds of satellite bandwidth! In July of 1980 Contel entered into a joint venture with Fairchild to take over ASC. In 1985 Contel bought out Fairchild's interest in the partnership and the company became Contel-ASC.

Contel-ASC entered into a contract with RCA Astro Electronics for three Series 3000 spacecraft and they built a TT&C facility outside of Atlanta. They also entered into a contract with NASA to launch the spacecraft on the Space Shuttle. ASC-1 was launched from the shuttle Discovery on Mission 51-1 on August 27, 1985. The plan was to launch ASC-2 from the shuttle, however, in the aftermath of the Challenger disaster President Reagan ruled that the shuttle would no longer launch commercial satellites. ASC-Contel sued NASA for the additional money it would cost to launch on an expendable launch vehicle and lost. ASC-2 was launched on a Delta 7925 launch vehicle on April 13, 1991. ASC-Contel cancelled their order for ASC-3 even though it was completely built. Subsequently, Rene Anselmo, who was starting up PanAmSat on a shoestring, made a deal with Astro to buy the $80 million bird for $45 million and renamed it PAS-1. He also made deal with Ariane to launch PAS-1 for 30% off on what was to be a no-payload test firing of the new Ariane 44LPH10. He only insured the launch for half of the value of PAS-1. It all worked and Rene had his satellite in orbit for literally pennies on the dollar. The only problem was that as designed for ASC, PAS-1's antenna pattern covered North America and Rene needed coverage in South America -- so they flew it upside down and it worked -- the North American Beam was now pointed at South America!

Chapter 12 - Expanding The Fleet

When F-3 was lost in December 1979, preliminary work had already started at Astro on two additional spacecraft. The project was immediately accelerated to get the first spacecraft completed and launched as soon as possible. The immediate crisis caused by the loss of F-3 was solved by the lease of Comstar D2. Americom now wanted to get the burgeoning cable business onto its own satellites and to stop paying over $800,000 per month to AT&T for the use of D-2. At the same time, Americom had learned some things in our three years experience with F-1 and F-2 and wanted to incorporate these improvements in the next spacecraft.

The new spacecraft were designated Satcom D and E during manufacture and would become F-3R (for F-3 replacement) and F-4 after launch. From the outside the spacecraft looked very much like F-1 and F-2, however, internally there were significant differences. Two tall thin nitrogen tanks were added on either side of the central cylinder of the spacecraft. These tanks contained the nitrogen pressurant for the fuel system that had originally been contained in the fuel tanks themselves. This allowed 125 more pounds of fuel to be put in the fuel tanks thus lengthening spacecraft life from 8 years to 10 years. With the huge investment in a spacecraft and its launch, the longer it lives, the greater the return on the investment.

Previously, I had described the attitude control problems with F-1 and F-2 caused by the plume from the North face thrusters 1, 2, 3 and 4 impinging on the solar array. For F-3R and F-4 thrusters 1 and 2 were canted slightly toward the West panel and thrusters 3 and 4 were canted slightly toward the East panel. This pointed the thruster plumes away from the solar array reducing the disturbance torques on the spacecraft. The net effect was to reduce the attitude disturbances and make exact array positioning for North/South maneuvers less critical.

The communication payload was also improved.   As in F-1 and F-2 the TWTA's were arranged in four banks of six tubes each.   Three of the banks contained 5.5 Watt TWTA's, a half watt more power than F-1 and F-2.   The fourth bank contained 8.5 Watt tubes.   Unlike F-1 & F-2, however, each of the four banks also contained one spare TWTA that could be switched in if one of the online tubes failed.   The individual transponders were so valuable (about $1million/year revenue), it only made sense to back them up. To power this larger payload, the solar array was increased from 75 sq ft and 800 Watts to 90 sq ft and 1000 Watts. The battery capacity was increased from 36 ampere-hours to 51 ampere-hours.

Of course all of these changes made the spacecraft heavier. F-1, F-2 and F-3 weighed 1953 pounds which was near the 2000 lb. maximum load that the Delta 3914 could inject in Geosynchronous Transfer Orbit (GTO).   F-3R and F-4 would weigh 2425 pounds so a launch vehicle with a higher throw weight to GTO would be required.   The solution was to replace the Delta Third Stage with the Payload Assistance Module D (for Delta).   The Thiokol PAM-D containing a Star 48B solid rocket motor was mounted as the third stage on the Delta 3910 launch vehicle.   The Star 48 indicates that the motor is 48-inches in diameter.   The PAM was originally developed to launch communication satellites out of the bay of the Space Shuttle but was adapted to replace the third stage of the Atlas (PAM-A) and the Delta (PAM-D launch vehicles.   It contains an electrically driven spin table that spins the spacecraft up to 60 RPM to stabilize it during Geosynchronous Transfer Orbit (GTO).   Both the F-3R and F-4 launches were successful, however, data analyzed post launch showed a problem with the PAM.   At the start of the Star48B burn the spacecraft was spinning symmetrically with no wobble. As the burn progressed, the PAM-D and the spacecraft started coning or wobbling around the spin axis.   Although thrust was reduced slightly because of off-axis burning, the spacecraft was successfully injected into GTO.   Post launch analysis revealed that the asymmetrical thrust was probably caused by part of the exhaust cone being burned away during firing.   If more had burned away, the coning could have become large enough for the PAM/Spacecraft to go into an uncontrollable tumble.   This near disaster prompted an intensive investigation of the whole exhaust cone manufacture process. The exhaust cone was constructed of built up layers of carbon-carbon composite material and epoxy that was then cured at high temperature. The failure was probably caused by an error in the long and complex process used to fabricate the nozzles. The process was revised and Americom successfully used the PAM on two shuttle launches in 1985 and 1986.

PAM-D in test stand prior to integration into the Delta launch vehicle. The exhaust cone is on the bottom of the rocket inside the test stand. - *NASA Photo*

The F-3R/F-4 spacecraft telemetry was changed from Pulse Amplitude Modulation (PAM) to Pulse Code Modulation (PCM) requiring new TT&C equipment in the ground station. The ground systems actually had to be expanded to handle the four new streams of PCM data from F-3R and F-4 while continuing to process four streams of PAM data from F-1 and F-2. We of course were expecting to continue to add more satellites so the system had to be further expandable in the future. The TT&C Data and Control computers were upgraded from HP-2100's to HP-1000's. These new computers were not only faster but had more memory and supported the Hewlett-Packard HPIB bus that allowed easy addition of many peripheral devices. A Perkin Elmer 8-32 minicomputer installed in the Vernon Earth Station replaced the leased UCC Univac 1108 in Dallas that we used for orbit determinations. Telos Computing of Santa Monica, CA ported over the Space 360 orbit software from the 1108 to the 8-32.

I had a Human Factors Engineering firm from Connecticut do a study of how Spacecraft Controllers worked at the TT&C Console. They observed controllers doing all of their functions such as commanding, ranging, monitoring maneuvers and collecting and plotting data. On the basis of this study, they proposed an ergonomically designed TT&C Console that incorporated the new equipment while making it easy and comfortable for the controllers to operate. The new consoles were installed at Vernon Valley and South Mountain.

Both the Vernon Valley and South Mountain earth stations were enlarged. At Vernon not only was space added to make up for that lost by the Tape Center installation. It also included communications expansion with a new console but also added a new Network Monitoring Center, a room for the Perkin Elmer 8/32 Computer, maintenance shops, lunch room and storage space. The South Mountain expansion was primarily for expanded communications and TT&C space and maintenance shop and storage facilities.

The launch team that assembled at Vernon Valley in the fall of 1981 for the launch of F-3R and F-4 was much different from the despirited team that had left in December 1979 after the loss of F-3. They were going to do two launches in quick succession and were enthusiastic to get on with it. The Mission Director was Larry Scholz who had a strong background in software and in spacecraft mission operations. He had been a member of the team that recovered the RCA-built Air Force DMSP-5D1 spacecraft that had tumbled out of control in space and been declared dead. It was a highly complex 6-month campaign in 1976 and 1977 to get the spacecraft under control, thaw it out and gradually get it pointed back at the earth. that resulted in the spacecraft being put back into service. Besides being a very smart guy, Larry was very much down to earth and had a great sense of humor.

Vernon Valley TT&C - L to R Duane McMillen, TT&C Manager, Spacecraft Controllers Harris Tyra, Gene Lampkin & Guy Crowther, Computer Technician "TP" Tubbs at the new TT&C console - *RCA Americom Photo*

With the expansion of the earth station, we now had much improved facilities at Vernon Valley. The Spacecraft Analysts now had a spacious office of their own. The Perkin-Elmer 8/32 computer was in a room of its own in the rear of the building. The TT&C area was now walled off from the noise of the communication area. The Tape Center had also been expanded and there was a new Network Monitoring Center (NMC) added.

Spacecraft Operations - L to R John Bailey, Analyst, Archie Miller, Manager, Christine Sharlow, Secretary, Steve Agid, John Schmidt and Fran Paulock, Analysts. - *RCA Americom Photo*

As in past launches, I intended to operate F-1 and F-2 from South Mountain during critical portions of the actual launch operations. Two analysts would be deployed to South Mountain to support F-1 and F-2 operations. These would be the last launches operated out of Vernon Valley. RCA was building the Astro Spacecraft Operations Center (ASOC) in the Astro plant in East Windsor, NJ. They were also building an earth station with TT&C facilities in Alpha, NJ and Guam, Marianas. The Alpha facility (sometimes called Carpentersville) was located in an abandoned quarry (for radio frequency shielding) near the Delaware River. The Guam station would get enough "back orbit" coverage so that Fucino and Carnarvon would not be required. We welcomed this development because, in the future, with four on-orbit spacecraft we would find it difficult to support a launch from Vernon Valley.

South Mountain TT&C - Front - Manager Ray Balon, S/C Controller Ed Ingram, Secretary Betty Vanderpool, S/C Controller Gary Scheller, - L to R Rear - Computer Tech Keith Hughes, S/C Controllers Phil Hamilton, Charles Boyd & John McAllister - *RCA Americom Photo*

Larry Scholz as Astro Mission Director for F3R and F4 was a true leader. He did an outstanding job of building and training his launch team. As a result the team was a group of enthusiastic and happy campers. Because the two launches were only two months apart, the F3R launch, spacecraft checkout and the then the F4 launch and checkout were basically one continuous operation. The launch rehearsals went smoothly as did the launches. F3R was

147

launched on November 20, 1981 and stationed at 131 degrees West Longitude.   F4 was launched on January 16, 1982 and stationed at 82 degrees West Longitude.   The only perturbation to operations was the extremely snowy winter.   We kept our snow removal crew mobilized and had several vans to ferry people back and forth from the station to the Playboy Hotel.

During the in-orbit bus tests, both spacecraft checked out well except for one major problem. One of the two command receivers (CMRs) on F3R had reduced sensitivity -- that it is took more uplink power than normal for commands to verify.   As a result of ranging tests through that receiver, Bob Youngblood calculated that one of the intermediate frequency (IF) amplifiers was probably shorted out.   At least it still worked even though it took more power.   As long as the other CMR worked, we would be OK. Unfortunately that wasn't to be.

After F3 failed in December 1979, Americom had leased transponders on AT&T's Comstar D2 for 10 of their F3 customers until F3R could be launched.   Americom held an "out of the hat" impartial drawing for D2 transponder assignments.   The table below shows the assignments.

Transponder	Customer
9	ESPN
11	Rainbow - Prime Time Network
13	Southern Satellite Systems (CNN)
15	Showtime
17	Home Box Office (HBO)
18	Home Box Office (HBO)
19	Satellite Communications Network
20	Spanish International Network
22	TCS - Pittsburgh Sports
23	United Satellite
24	National Christian Network

The plan now was to transfer these customers from Comstar D2 to F3R.   It would not be easy because there were now literally hundreds of cable system antennas pointed at Comstar D2 at 95 degrees West Longitude. These antennas would have to be repointed to F3R at 131 degrees

West Longitude at the other end of the North American equatorial arc. Some of the larger antennas had mechanical limitations requiring heavy equipment to move the jack screw assemblies. Obviously all of the antennas could not be moved simultaneously so Americom developed a transition plan that gave the cable operators several weeks to move their antennas. Americom, using their own earth station facilities, down-linked all of the TV signals from Comstar D2 and retransmitted them up to F3R. Now the same programming was available on both satellites so the operators could transition at their convenience. Also if they had problems moving their antennas, they could always go back temporarily to D2. There were also customers on F-1 who were transitioning to F-3R. Americom made similar double-feed arrangements for the F-1 customers.

After all of the transitions were made, the following table shows how F-3R traffic looked as 1982 drew to a close

F-3R Customers Late 1982			
ODD V-Down	Customer	EVEN H- Down	Customer
1	Nickelodeon	2	PTL - Religious
3	United Video - WGN	4	Spotlight
5	The Movie Channel	6	Southern Satellite - WTBS
7	ESPN	8	Christian Broadcast Network
9	USA Cable Network	10	Showtime West
11	MTV	12	Showtime East
13	HBO West	14	Southern Satellite - CNN
15	Warner - AMEX	16	Showtime & Appalachian
17	Showtime	18	Reuters/EWTN
19	Occasional	20	Cinemax East
21	Landmark - Weather Channel	22	HBO & Modern Sat Network
23	Cinemax West	24	HBO East

By late 1981, just before F4 was launched, the CATV business was booming and there was an enormous demand for satellite transponders. Americom decided that the demand was great enough that instead of leasing the transponders at a fixed price, they would auction off some of

149

the F4 transponders before it was launched. RCA's position was that because the company invests time and money and takes the risk, it should reap the rewards, not transponder speculators. In the past speculators won leases through lotteries and then sold the transponder rights for millions in profit. An auction "is the free marketplace in purest form" said Andy Inglis, President RCA Americom. Past lotteries were "purely random" and had " no rational basis". "The company that has made the investment and has taken the risk of providing facilities' should reap the financial rewards of the risk" Andy said.

Eight of the 24 transponders had already been sold to customers who had originally had been scheduled for the ill-fated Satcom-3 that was lost on launch. Two more were preemptible transponders used for backup and couldn't be sold long-term. Americom advertised that seven F4 transponders would be leased for seven years at auction to be held on November 19, 1981 at the Manhattan galleries of Sotheby Parke Bernet.   Standing in front of a golden colored 1/4-scale model of the F-4 spacecraft, Sotheby President John L. Marion opened the bidding. Bidding quickly went from $50,000 to $14.4 million by a previously unknown company called Transponder Leasing Corp.   It was expected that they would sell the lease to a programming company. Billy H. Batts, a religious broadcaster, got the next one for 14.1 million. The rest went to Warner-Amex for $13.7 million, RCTV (a Spanish Language Programmer) for $13.5 million, HBO for 12.5 million, Intercity Broadcasting for $10.7 million and UTV Cable Net for $11.2 million. Altogether the bids for the seven transponders for seven years totaled $90.1 million.   That was about $20 million more than Americom would have made if the transponders had been leased under their previously announced F-4 rate schedule.

Of course, it couldn't be that easy.   There were already 17 petitions before the FCC objecting to the auction and asking the FCC to overturn it. The main argument was that selling transponders to the highest bidder was "unjustly discriminatory" and inflated the prices. The FCC had until January 15 (3 days after the scheduled F-4 launch) to disallow the auction and prohibit all future auctions. In January, the FCC ruled against the auction primarily because it resulted in different prices for what was effectively the same piece of equipment.

RCA Americom then petitioned the FCC to allow them to lease the transponders with pricing set on the basis of demand rather than their actual cost. Americom set the price for the seven transponders at $13 million each based on the average of the winning bids at the November auction. If there were more than seven bids, the winners would be selected by lottery. The FCC approved and the lottery was set for March 29, 1982. The FCC also said that they were launching an investigation of transponder leasing and if they decided that the lottery prices were too high, Americom would have to give rebates. The ruling was immediately appealed by a number of parties on the basis that if the price was excessive, some could not afford to bid and would be left out if ultimately the FCC dropped the price to one they could afford. The lottery was put on hold pending the decision.

On March 31, a Federal Appeals Court declined to block the regulatory order and the lottery was scheduled for April 6, 1982. To be included in the drawing, an applicant had to submit an order for an F-4 transponder between 9AM and 3PM to the lobby of the Americom headquarters on College Road in Princeton along with a certified or cashiers check for $13 million. The lottery, if required, would be conducted at 3:15PM in a building in the David Sarnoff Research Center in Princeton supervised by the accounting firm Deloite-Touche. The lottery was not open to the public.

Unfortunately, the weather gods didn't seem to smile on this enterprise. On Tuesday April 6, 1982, the day of the lottery, it started to snow in the early morning and quickly turned into a swirling blizzard with wind gusts as high as 60 miles an hour. At some point during the day, there was thunder and lightening as a cold front passed through. In Princeton over 10-inches of snow fell and then was piled into huge drifts by the swirling winds. Not a great day for travel in central New Jersey. Three intrepid bidders showed up -- but they were reluctant to hand in their checks if they were going to be the only bidders.

So now Americom had to go out and sell the transponders. Remember, the lottery was only for seven of the twenty-four transponders on F-4. One of the reasons for that was that eight of the F-4 transponders had already been sold. HBO, for instance, had the first four transponders on F-4 and was the prime reason that there was so much interest in F-4 by other program suppliers. Americom felt that the HBO national service anchored the bird, guaranteed that all of the other transponders would sell and locked in profitability. Unfortunately, however, at about the time of the lottery HBO decided that they did not want the space that they had already bought on F-4. They said that F-4 did not have enough signal power and that they would wait for a Ku Band bird which would be the next big thing. HBO, of course, had a contract with Americom that included termination penalties. After some negotiation, Americom entered into a deal with HBO that released them from their F-4 contract in exchange for their partnership in Crimson Satellite Associates, a two satellite medium power Ku band venture (more of that later). The value of F-4 transponders dropped like a rock. Some of those already signed up for transponders on F-4 complained that without HBO, F-4 did not have any national cable reach. Some, led by Cablevision, threatened to move to other birds. Bill Berman, one of Americom's marketing guys came up with a solution to the problem -- give away dish antennas to to the largest cable systems to assure that there would be a customer base looking at F-4. Although, at that time, there were about 10,000 cable systems in the US, 2500 cable systems actually contained about 80% of the cable subscribers -- these were the systems that got the free dishes.

Based on this give away program, Berman then proposed a program to sign up regional sports programmers who were then popping up all over the US. Cablevision was already developing regional news and sports programming to go on its Madison Square Garden channel on F-4. Berman's idea was to provide the antenna base (and thus customers) for these new sports networks and in turn they collectively formed a mosaic that covered most of the populated areas of the US. This mosaic would attract national programmers and enhance the value (and revenue to be gained) from F-4. The two concepts, free dishes and regional sports networks worked hand in hand. Ultimately F-4 carried Madison Square Garden, Sportsvision (Chicago), Home Sports Entertainment Network, Sportschannel New York, New England Sports Network, Florida Sports Channel, Prime Ticket (So. California), Home Team Sports (DC & Maryland) and Sports Channel New England. The remaining transponders were occupied by Playboy, Bravo, American Movie Classics, WPIX Superstation, a number of shopping channels and several religious broadcasters. HBO's departure left F4 in a rather perilous business position and led Americom to the antenna seeding and mosaic strategies that were able to create a national bird without a national program anchor.

But what made HBO quit their deal with Americom? Was it really that F-4 wasn't powerful enough and that they were waiting for K-Band? It turned out that there was a new (and very competitive) kid on the block -- Hughes Communications. Up until this point Americom held the lion's share of the cable TV business and the bureaucratic AT&T and the bumbling Westar split the rest. Hughes had been building satellites for everyone else, (Anik, Westar, AT&T) and now they were planning to launch three Galaxy spacecraft for a communication business of their own.

Hughes had hired Clay Whitehead ,who had been the director of the White House Office of Telecommunication Policy in the Nixon administration, to head Hughes Communications. Hughes announced that they would launch a new more powerful bird and would charge low transponder prices that Americom would not match. Americom felt that they had the premier birds at the premier positions and did not want to weaken their price structure. Whitehead had started by signing up potential customers before Hughes even began building Galaxy 1. As a result, Hughes had a backlog of customers lined up by the time Galaxy 1 (G-1) was launched on June 28, 1983. Galaxy's business plan was to dedicate their satellites to cable TV -- no PLC business. The Galaxy birds had 9-Watt transponders as compared with the 8.5-Watt on Americom's F-3R and 5-Watts on F-4. Hughes was advertising Galaxy-1 as the "hot bird".

That was what HBO was talking about when they abandoned Americom's F-4 -- but was it the real reason? My understanding was that Hughes knew that to dominate the cable TV business they needed HBO to anchor Galaxy-1. Industry lore has it that Hughes went to HBO and asked what it would take to move them to Galaxy -- and that was the deal that Hughes gave them. Americom would not match the HBO deal offered by Hughes. By October, 1983, nineteen of the twenty-four G-1 transponders were filled with cable programming. HBO had six transponders, Group W/Westinghouse had four, Viacom two, Turner two and Times Mirror two.

For Americom it was a new much more competitive world than it was with only AT&T and Westar to worry about. But just as they had adjusted to the loss of HBO on F-4, they also rose to the threat of the new competition. When Americom designed Satcom C-1 and C-3 to replace F-3R and F-4 they insured that their signals would be stronger than the competition. Americom elected to use 10-Watt TWTA's for C-1 and C-3 and shaped the transmit beam to increase the EIRP levels in the lower 48 states to at least 2dB greater than Galaxy 1. These would be the new "hot birds" in the cable marketplace.

Chapter 13 - Advanced Satcom

As the business grew and the original satellites aged, there was a need for replacement spacecraft. F-1 and F-2 were past their half-life, had a number of on-board failures and needed to be replaced. The original Alaskan satellite, F-2, had demonstrated how well satellite communications worked in Alaska's rugged terrain. The new Astro 3000 spacecraft evolved from the original design but incorporated lessons learned from the earlier birds and new technology developed over the intervening years. RCA Alascom required a new spacecraft completely dedicated to Alaska and designed for Alaskan communication service. Alascom also required access to another spacecraft that could backup their dedicated bird in the event of a spacecraft failure. Three Advanced Satcom spacecraft would be built to fulfill these requirements. F-5 would be dedicated to Alaskan service and would cover Continental US (CONUS), Hawaii and Alaska on the vertical polarization and Alaska and Hawaii only (no CONUS) on the horizontal polarization. F-1R would cover Alaska, Hawaii and CONUS in both polarizations and be switchable to Alaska and Hawaii only on horizontal to backup F-5 if required. F2R provided CONUS only coverage on both polarizations.

Americom sold transponders at three different levels. Protected Service (highest cost) guaranteed that if your transponder, or the spacecraft it was on, failed, you would be restored in another transponder or spacecraft. Unprotected Service (median cost) guaranteed that although you would not be protected, you also would not be preempted to protect someone else. Preemptible Service (lowest cost) meant that you could lose your transponder to protect someone else. F-5 would serve Alaska and have 22 protected transponders and 2 preemptible. F-2R would serve the U.S. lower 48 states and Hawaii and have 22 protected transponders and 2 preemptible. F-1R would serve as the protection spacecraft for both F-5 and F-2R and so would have 2 protected transponders and 22 preemptible. The two preemtible transponders on F-5 and F-2R would provide protection for individual transponder failures on those spacecraft.

Advanced Satcom spacecraft showing the two antenna reflectors and the offset feed.   The solar array shafts are mounted in the center of the spacecraft box. This is an artists conception -- while the spacecraft and earth relative sizes are reasonably accurate, it appears that the bird has lost lock on the earth, is spinning about the pitch axis and nutating! - *RCA Astro-Electronics Division Photo*

The four petal antenna with feed horns in the center performed very well on the first four Satcom's. It's design resulted in a domed elliptical shaped beam with the ends of the ellipse over Alaska and Florida (see the illustration on the cover page).   It, however, wasted power in uninhabited areas in Canada and had weaker signals in coastal Southern California and "Down East" Maine..   The Advanced Satcom antenna had two overlapping 7-foot diameter reflectors (horizontal & vertical) with an offset RF feed to minimize "shadowing" of the dishes.   The multiple horn feed shaped the beam so that it concentrated power in the desired service area with minimum "spillover" into unpopulated areas.

Prior to the Advanced Satcom, virtually all government or commercial communication spacecraft used Traveling Wave Tube Amplifiers (TWTA) for their communication transmitters. Satcoms 1, 2, 3R and 4 used 5 Watt TWTA's with the exception of 3R that had one bank of six 8.5 Watt TWTA's. The Advanced Satcoms used 8.5 Watt Solid State Power Amplifiers (SSPA). There were seven SSPA's for each bank of six channels -- six operating amplifiers plus a spare. The power supplies for the SSPA were separate units each powering multiple SSPA's through a fairly complex switching scheme. The primary advantage of the SSPA's was that their linear amplifier performance reduced Intermodulation Distortion and thus allowed higher traffic loading (more circuits) in each transponder.

A number of things were done to reduce attitude disturbances caused by the North face thruster plumes impinging on the Solar Array during North/South maneuvers. In the first four Satcom's, the North and South solar arrays were mounted on a single shaft that went through the spacecraft box structure. Because the Apogee Kick Motor (AKM) occupied the center of that box, the array shaft had to be located close to the spacecraft's earth face rather than going right through the center of the North and South panels. On the Advanced Satcoms, each array had a separate shaft and drive motor and thus could be located in the center of the North and South panels. The North and South array positions could be controlled separately which also allowed them to be adjusted individually to minimize the effects of solar pressure on spacecraft attitude. The North face thrusters 1, 2, 3 and 4 were moved away from the center and out to the four corners of the North panel.

A major innovation on the Advanced Satcom's was the addition of Electrically Heated Thrusters (EHT). EHT's were not yet available when F-5 was under construction so only F1R and F2R received them. EHT's consist of a conventional mono-propellant 0.2 pound hydrazine thruster with an electric incandescent heater that heats and adds energy to the thrust gases. The EHT, also called a Resistojet, was invented by RCA Astro scientist Yvonne Brill and represented a monumental advance in the field of single-propellant rockets. EHT's, which can only operate in the vacuum of space, are much more fuel efficient than mono-propellant

157

thrusters and increased the life of F-1R and F-2R,

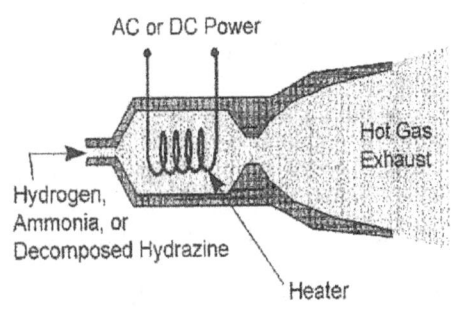
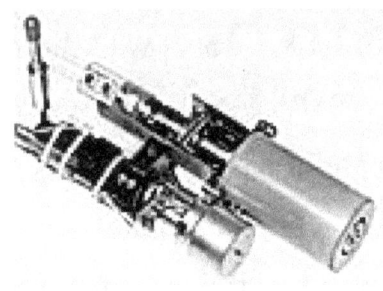

The schematic drawing above shows how decomposed hydrazine from a conventional 0.2lb thruster is heated by the EHT's electric heater. The photo shows an actual EHT thruster assembly. The smaller cylindrical assembly is the 0.2lb thruster whose decomposed hydrazine output goes through the small black tube between the cylindrical units into the EHT (larger cylinder) and the heated gases are exhausted to the right. - *Public Domain*

As noted above, the North Face thrusters 1, 2, 3 and 4 were moved out to the corners of the panel. On F-1R and F-2R, the EHT thrusters, 13, 14, 15 and 16 were arranged around the solar array shaft in the center of the panel.   For North/South maneuvers two EHT's (13 & 15 or 14 & 16) provided thrust in a continuous burn.   The on-board computer fired thruster 1, 2, 3 or 4 at 1/4-second intervals as required to maintain spacecraft attitude.

The Advanced Satcom's were originally planned to be launched on the Space Shuttle, however, the shuttle development program had fallen far behind schedule and RCA made the decision to launch all three spacecraft on Delta launch vehicles.   In the time since the F-4 launch, Astro had built the Astro Space Operations Center (ASOC) in their plant at Locust Corners in East Windsor, NJ.   They had also built two TT&C stations; one in an abandoned quarry in Alpha, NJ near the Delaware River and another on the island of Guam in the Western Pacific Ocean. Although these two stations did not achieve quite as much orbital coverage as when Astro used the Fucino, Italy and Carnarvon, Western Australia stations for launch support, their studies showed that it was sufficient for safe reliable launch support.   I, for one, was very happy to see this, because with the level of activity at Vernon Valley and South Mountain supporting four spacecraft plus other communication activities, it would have been very difficult to support launches. Vernon Valley and South Mountain would, however, provide ranging support and forward telemetry data to ASOC during transfer orbit. During launch and early on-orbit operations, ASOC would forward telemetry data into the Americom system so that we could monitor all operations in real time.

The F-1R spacecraft being lifted onto the launch vehicle adapter ring to verify the fit of the interface. Note the soft materials on the floor to prevent damage to the Delta launch vehicle side of the interface. - *RCA Astro-Electronics Division Photo*

F-5 was launched on October 28, 1982 on a Delta 3924 launch vehicle. The Apogee Kick Motor was fired successfully and the spacecraft drifted on to its station at 143 degrees West Longitude. The spacecraft bus testing showed the spacecraft to be operating within nominal parameters. Some of the communications system testing was done from Alaska to verify the transmit and receive beam coverages.

The F1R launch was scheduled for six months after F-5 in April of 1983.   Because of my responsibilities and the key role of the Vernon Valley and South Mountain stations in earlier launches, I was never able to go to the Cape to participate in a launch.   I was very pleased when my boss Joe Schwarze called me and said that I was invited to Cape Canaveral for the F-1R launch scheduled for April 11, 1983   Catherine and I flew to Melbourne and joined all of the other Americom and Astro folks at the Cocoa Beach Holiday Inn.

The RCA Americom F-1R launch team. L to R Joe Schwarze, Director Space Systems, Archie Miller, Manager, Spacecraft Operations, Bill Palme, Manager, Launch Vehicle Integration, Jim Tietjen, President RCA Americom and John Christopher, VP Technical Operations. - *RCA Americom Photo*

On Monday, Joe Schwarze and I attended the T-3 Flight Readiness Review at the Delta Program Office.   This review is an update of activities and is conducted to review the test and checkout data to verify that the launch vehicle and spacecraft are ready for countdown and launch. Upon completion of this meeting, authorization is given to proceed with the loading of the Delta second-stage propellants. This review also assesses the readiness of the range to support launch and provides a predicted weather status. The investigation and resolution of each problem encountered during launch vehicle construction were covered in detail.

Delta 3924 launch vehicle carrying F-1R on the launch pad next to the Fixed Umbilical Tower (FUT)on the morning of April 11, 1983 about eight hours before launch. The Mobile Service Tower (MST) has been moved away and the RCA Spacecraft Checkout Station in Hanger AE is communicating with F-1R by telemetry and command radio links. - *Archie Miller Photo*

Problems with other Deltas that could possibly impact our launch vehicle were also described and the investigation to determine relevance to our program described. The F1R launch vehicle used a new system to eject the solid booster motors more quickly and it was described in detail. The old system used large compressed dished washers on the explosive bolts that pushed motor away as the compression was released. The new system used hydraulic/pneumatic actuators that worked faster and with more energy. Getting rid of the motor weight faster gave us another 60 pounds of throw weight to Geosynchronous Transfer Orbit.

Although the launch vehicle was checking out OK, the same could not be said of the spacecraft. F-1R was in the Delta Spin Facility while the Apogee Kick Motor (AKM) rocket was being installed. As the AKM was being inserted into the central tube of the spacecraft, they found an interference problem between the AKM and a section of the communications system waveguide. The fix to the problem was relatively simple, however, it caused a day delay pushing the launch to Friday. On Friday high winds aloft scrubbed the launch. Saturday and Sunday there were high winds, fog and thunderstorms. Monday dawned calm and clear with blue skies and sunshine.

Bill Palme, Americom Launch Vehicle Integration Manager, in the Mission Director Center (MDC) in what was then called Hanger AE at Cape Canaveral. - RCA Astro *Photo*

Launch operations and overall mission activities were monitored by the F1R Mission Director (MD) and the supporting mission management team in the Mission Director Center (MDC) in Hanger AE, where the team was kept informed of launch vehicle, spacecraft, and tracking network flight readiness. Each of the RCA Astro and Americom personnel plus Delta and Range persons were assigned console positions in the MDC. A short distance away in the AE White Room, RCA Astro Engineering personnel were manning the Spacecraft Checkout Station monitoring F-1R telemetry data and sending commands as necessary. As we manned our console positions, a few hours before launch, the F-1R batteries were being charged to "top them off" before launch.   In the Delta Blockhouse located about 1000 feet from the launch pad, the Douglas engineers were verifying the launch vehicle status.   At our console positions we could see spacecraft telemetry and could hear the conversations on the the launch network --we were really just observers with no responsibilities for the launch.   We had, in fact, agreed that as soon as the Delta lifted off, we would all go outside to watch it climb out.

The final countdown starts at 150 minutes before launch (T-150).   All personnel not involved in the count are cleared from the launch area -- the warning horn sounds.   At T-146 minutes the first and second stages helium and nitrogen pressurization begins.   A T-140 the launch vehicle guidance system is turned on and at T-130 the first stage fueling begins.   At T-95 there is a weather briefing.   KSC meteorologists give a detailed weather forecast, including winds aloft predictions, for the next two hours.   If the weather report is favorable, at T-87 the launch team is polled for readiness to begin loading liquid oxygen (LOX), at T-80 the blast danger area is cleared and at T-75 liquid oxygen loading begins.   At T-60 the vehicle C-Band radar transponder beacon checks begin.   This is a critical check because the data from this radar defines the launch vehicle trajectory as it climbs out from the Cape and proceeds down range. At T-30 the Delta controllers perform engine gimbal steering checks.   They send commands to slew the main engine nozzles to assure that they are free and operating correctly.   At T-20 there is a built-in 20 minute hold.   During this hold there is a weather update by the meteorologists. All of the members of the launch team are polled to confirm that they are ready to launch. When the count picks up again at T-20, the activity accelerates.   The liquid oxygen tank is topped off to 95 percent.   The helium and nitrogen pressurization systems are topped off and the Range Safety Command Carrier is on.   This radio carrier will be used to command "destruct" in case the launch vehicle veers off course.   At T-14 checks are run on the Range Safety Command Receiver to assure that it will execute destruct commands.   At T-10 the first stage fuel tank is pressurized and at T-6, liquid oxygen is topped off to 99 percent.   At T-6 there is another weather check to verify "Go for launch".   At T-5 the F-1R spacecraft is switched to its internal power (batteries) and the Launch Enable switch is moved to ON.   At T-4 there is a built-in 10 minute hold.   At the end of this hold the Launch Conductor receives "go" to release from hold. From here the count will proceed non-stop to liftoff.   When the count picks up all of the launch vehicle and Range Safety pyrotechnic devices are armed and the launch vehicle goes on to internal power. At T-3 minutes the Astro Mission Director reports that F-1R is go for launch. At T-2 minutes the liquid oxygen tank is raised to flight pressure. At T-80 seconds they top off the first stage liquid oxygen to 100 percent. At T-70 seconds the KSC range facilities report go for launch. At T-60 seconds verify Launch Enable switch ON. T-30 seconds the liquid oxygen fill and drain valve is closed. T-7 seconds Go for main engine start. T-2.5 seconds Main engine start. T-0 LIFTOFF at 5:39PM EST!

Instantly there is a loud low frequency rumble that shakes Hanger AE. We all pull off our head phones and head for the door and into the parking lot. I go to my car and get my camera and start shooting as the Delta climbs away from the Cape. The noise is hard to describe -- I am feeling it as much as hearing it. It is a beautiful clear day with unlimited visibility. As the Delta continues to climb we see the first three Castor solids jettisoned and go flying away. It is a picture perfect launch. As it climbs out of sight the six other Castor solids are jettisoned and the first stage separates and falls away. The second stage ignites and continues the vehicle southeastward. When the second stage burn is complete and the stage falls away, the vehicle enters a coast phase -- at this point the vehicle is about 100 miles above the earth. As the vehicle crosses the equator, the third stage solid propellant motor ignites and puts the spacecraft into the highly eccentric Geosynchronous Transfer Orbit (GTO).

The Delta 3924 carrying F-1R climbing out of Cape Canaveral and starting down range. - *Archie Miller Photo.*

After we got the initial tracking data from KSC it turned out that we had gotten a rather "hot" ride and were a bit higher than we wanted to be.   Normally we would track the spacecraft for three days and fire the AKM on the seventh apogee which would occur on Thursday.   Because of the high orbit, F-1R would be 50-degrees further west and drifting away from its station at 139 degrees West.   AKM fire was postponed until the ninth apogee on Friday when F-1R would be in a much better position.   AKM fire was successful and F-1R slowly drifted to its station at 139-degrees West Longitude.

As noted earlier, the design of the North face thrusters and the position of the solar array allowed North/South maneuvers independent of the solar array position.   Also the primary

thrust for these maneuvers was provided by the Electrically Heated Thrusters (EHT)
Thrusters 1 through 4 were used primarily for attitude control, however, their thrust (small as it may be) contributed to orbit plane change. The way it worked was that two of the four EHT's burned continuously to provide the inclination change. Thrusters 1, 2, 3 and 4 at the four corners of the spacecraft North face fired pulses as necessary to maintain spacecraft attitude in roll and yaw. They fired short pulses singly and in pairs according to an algorithm in the North South logic in the Attitude Control Electronics (ACE). Data from the Earth Sensor Assembly (ESA) provided Roll error data to the ACE. The Yaw Gyro provided Yaw Error data to the ACE. While the EHT's were burning, thrusters 1, 2, 3 and 4 were controlling in Coarse Mode. After the EHT's were turned off thrusters 1, 2, 3 and 4 fired much smaller pulses in Fine Mode for a short time to settle down any remaining disturbances and then the ACE shut down the maneuver. During in-orbit testing, a number of short North/Souths were performed using all combinations of EHT and attitude control thrusters.

We did not have to do a "real" North/South for some time because the AKM fire resulted in the spacecraft inclination being on the "high" side. We had to wait until F-1R was well within the box before attempting to correct inclination. Our first North/South (N/S) on June 28 started smoothly. Serendipitously, the solar arrays were approaching the dawn (East Side) position which was the approximate array position for N/S maneuvers on our earlier birds. All of a sudden, F-1R experienced a large roll error and the ACE aborted the maneuver automatically. The spacecraft was still locked on the earth but nutating heavily. We sent command lists to fire thrusters to kill the nutation and then had to do a number of thruster roll controls as an apparent large yaw error translated into roll. After we had F-1R settled down, we gathered all the maneuver data and tried to figure out what had happened. Evidently there was an instantaneous event that caused a large attitude disturbance. We also sent the data to Americom Engineering and to Astro. It wasn't long before we got a call from Astro summoning us to a meeting at the Astro plant the following day. We couldn't figure out what had happened but Astro had.

The meeting was chaired by Steve Fox, one of Astro's lead attitude control engineers. In attendance in addition to Americom technical personnel were a large number of Astro engineers including Warren Manger, Astro's Chief Engineer. This undoubtedly was a big deal. Steve went through the maneuver data up to the large transients at the time of the abort. Steve pointed out that the spacecraft nutation period changed abruptly at the abort time. He said that this was because the north solar array had folded at the upper hinge! Holy mackerel -- what kind of force had caused that? Evidently after the maneuver aborted, the springs in the array hinges caused the array to go back to its normal position. The attitude people thought that it was an artifact of the solar array position and would not occur again as long as we did not have the array in the dawn or dusk position for the maneuver. Evidently, because thrusters 1- 4 had been moved to the four corners of the North face, the array was being excited by the pulses, particularly at the dawn and dusk positions. It also appeared that the resultant disturbances were primarily in yaw. After the meeting, I discussed the situation with Joe Elko, Americom's Spacecraft Engineering Manager. He and I agreed that we needed to review all the data very carefully to see if there were any others clues as to what happened. We were also agreed that the next F-1R North/South maneuver would be an "all hands evolution" with all of the technical and engineering people at Vernon Valley just in case something happened.

The next North/South was scheduled for Wednesday July 6th. Joe Elko and most of his engineers were at Vernon Valley for the maneuver. The arrays were not in the dusk or dawn position -- in fact they were at the midnight position. The maneuver started at about 3:45PM local time. The maneuver started smoothly and everything seemed to be working well. Thrusters 1, 2, 3 and 4 were pulsing regularly and were keeping Roll Error and Yaw Error well within limits. After a while Yaw Error started increasing occasionally. At 15 minutes and 3 seconds into the burn something suddenly happened. The Yaw Error increased suddenly and went to maximum, Roll Error increased rapidly and the spacecraft lost Pitch lock on the earth.

As the Yaw and Roll Errors increased the ACE automatically shut down all of the thrusters, however, by this time the spacecraft was very badly disturbed. Because of problems with earlier spacecraft we had experienced loss of pitch lock before and had experience correcting large attitude disturbances but we had never seen anything as bad as this. One of the analysts reported that at the time of the incident, the Thruster Data File showed that thrusters 1, 2 and 3 had stopped firing and thruster 4 had burned continuously for 12-seconds! No wonder we were blown off the world. It appeared that F-1R had Roll Errors exceeding 45 degrees, was nutating about 45 degrees and was spinning around the Pitch axis. These were all estimates, of course, because numbers this large were way outside the range of the Roll Error and Yaw Error telemetry.

We immediately executed our recovery procedures -- we sent the PITCHOFF command list that configured the spacecraft for recovery, put the beacons in High Power mode and put battery chargers in High Charge mode. We knew that the first thing that we had to do was to remove the nutation -- from the look of things, it wasn't going to be easy. We had command lists called DENUTES set up to fire thrusters singly or in pairs. The thrusters had to be fired precisely at the peak of the nutation cycle. We were having a hard time doing that. The spacecraft was so far over that we were intermittently losing telemetry and command capability because even the spacecraft omni-directional antenna was sometimes pointing away from the earth. When we lost the signal at Vernon Valley, South Mountain usually had contact and vice versa. As a result we commanded DENUTES from whichever station was in contact at the moment.

About 20 minutes after we lost lock on the earth, Dave Stewart, the power engineer in Joe Elko's engineering group came to me and advised "you are going to have to turn off the payload". He said that because of the spacecraft attitude and the fact that it was spinning, it was not getting enough sun on the solar array to charge the batteries and power the payload. If we left the transponders on we would discharge the batteries to a dangerously low level. The next time Vernon Valley had a good command path to the spacecraft we commanded all of the transponders off. The payload generated a lot of heat and with it off the spacecraft was starting to get cold. The thermal guys were carefully monitoring spacecraft temperatures to make sure that we weren't freezing something particularly fuel lines. Now that we were in a better power situation with the payload off, we could spare some power to turn on heaters if required -- and we did. I thanked my lucky stars that we had all this highly qualified help available to worry about all the spacecraft systems. My analysts and I were completely occupied trying to figure out how to get the spacecraft right side up. When it became apparent that recovery would take a while, the Transponder Restoration Plan for F-1R was executed and our customers were moved to other Americom spacecraft.

Obviously, we had a very difficult recovery ahead of us. Although we in Spacecraft Operations prided ourselves on our ability to recover badly disturbed spacecraft, we had never seen anything quite like this. Joe Elko had called Astro and asked them to send a couple of their best attitude control guys up to Vernon Valley. In the meantime, we kept sending DENUTE command lists from Vernon Valley and South Mountain. The fact of the matter, however, was that our command lists were just nibbling at the problem. They were not firing long enough pulses to take out the large errors we currently had in the bird -- and we were reluctant to increase pulsewidths for a number of reasons. One was, of course, inexperience with this level of catastrophe and the other was the difficulty in interpreting the attitude errors from the earth sensor and the yaw gyro data with the spacecraft this far off the earth. There were great opportunities to make mistakes. But help was on the way from Astro.

While all of this was going on, my people were plotting telemetry data from the actual incident itself and distributing it to the Americom and Astro engineers. We had to find out what happened -- at the moment it was still a mystery. Why out of the blue, when everything had been working perfectly, did the ACE suddenly go beserk and fire thruster 4 for 12 seconds? After much poring over the data, the answer was not immediately obvious.

We continued all Wednesday night firing thrusters and gradually reducing nutation but not nearly enough. Thursday morning Carl Huber and another Astro attitude guy arrived and started assessing the situation. They built some new command lists to use north face thrusters with much longer thruster firing times.   One of the major problems in a situation like this is that every quarter-orbit roll error and yaw error interchange. As the spacecraft goes around the earth, roll translates into yaw every six hours and then translates back to roll six hours later. When the yaw error is large, the roll error is small and can be seen in the earth sensor.   When the roll error is large, the yaw error is small and can be seen in the yaw gyro.   So the strategy for firing thrusters changes as the spacecraft goes around the orbit.   Shortly before midnight Thursday the Astro guys had pretty much removed all of the nutation.   Roll was at maximum (probably over 40 degrees) and yaw was at minimum.   It would be six hours before roll came down so that it was within the 2.25 degree range of the Earth Sensor and we could get it locked up on the earth.   The Astro guys went back to the Playboy Hotel to get some sleep.

I had gotten a nap in the afternoon and planned to stay through the night until we got F-1R locked up on the earth.   Now that the spacecraft wasn't nutating, we were not having any problems receiving telemetry or sending commands.   We had the telemetry receiver signal strength (AGC) displayed on a strip chart recorder.   As time went by and the spacecraft roll error decreased, the strip chart plot showed signal strength gradually increasing.   We were receiving telemetry from the spacecraft omni-directional antenna -- the high gain communications antenna was not pointed at the earth.   About 3AM as roll error continued to decrease we began to see short intervals of much higher signal levels as the communication antenna beam swept past the earth.   Remember, in addition to the roll and yaw errors, F-1R was also spinning around the pitch axis.   About 3:30 AM I called the Playboy and suggested the Astro guys get into the station because the roll was coming down pretty fast.

Around 4AM the F1R Earth Sensor began to sweep over the earth and we had roll error information.   We started sending a barrage of command lists firing thrusters to correct roll

171

errors, to remove yaw and adjust momentum to slow the spin rate. Finally at about 6AM Friday morning F-1R finally locked up on the earth. It had been 38 hours since F-1R had been "blown off the earth". We were all exhausted at this point but still had a lot of work to do. We systematically went through all of the spacecraft systems and confirmed that they were configured correctly for normal on-station operations. One of our first priorities was to get the batteries completely recharged. The solar arrays were adjusted to be sure that they were square on the sun. The payload was not turned on. The communications engineers wanted to do an instrumented turn-on of each transponder to verify normal turn on signatures. They were also going to do quick performance tests of all transponders so that they could assure customers that communications performance had not been impaired by the incident and the subsequent emergency shutdown.

We also commenced ranging on F-1R so that we could get an orbit to find out where the bird really was after all the thruster firing. We also went through thruster data files and all of the commands sent to F-1R to determine how many seconds each thruster had been fired to calculate fuel usage. On Saturday morning, Spacecraft Analyst Steve Agid processed the range files and ran an orbit determination. On the basis of that orbit he planned an emergency East/West maneuver to keep F-1R from going out of its assigned stationkeeping box -- all of that thruster firing had pushed it westward.

On Sunday, we did some North/South maneuver tests running short maneuvers with different thruster combinations. Astro wanted to evaluate the performance of different thrusters and their effect on spacecraft attitude. The following week we continued doing tests on F-1R in support of Astro's investigation. My analysts and I were reworking our F-1R North/South procedure particularly the maneuver termination portion. The communication engineers were at Vernon Valley to turn on each of the transponders and test them. Americom management was in no hurry to put F-1R back into service. We had restored our customers to other spacecraft and now needed to find out precisely what had happened and how to prevent a recurrence.

After extensive review of all of the data, Astro finally determined what had caused the F-1R catastrophe -- we had discovered the spacecraft's resonant frequency! Determining the real resonant frequency of a spacecraft prior to launch is almost impossible. First of all, the solar array and its supporting arms cannot support their own weight in a 1g environment -- so it is impossible to assemble or test the whole structure on the ground. The mechanical engineers had calculated that the resonant frequency of the assembled spacecraft structure was about 0.3 Hz -- it turned out to be 0.5 Hz. When doing an EHT maneuver with thrusters 1, 2, 3 and 4 in Coarse mode, the ACE is clocking thruster fire at a 0.25 Hz rate. The second harmonic of 0.25Hz is 0.5Hz that equals the resonant frequency of the spacecraft structure. The net effect was that the whole structure and particularly the solar arrays started to vibrate at the half-Hertz rate. That is why the solar array folded up in our first maneuver. In normal operation, the yaw gyro drives an up & down counter that provides the digital output to the Attitude Control Electronics (ACE). During the fateful maneuver of July 6, the vibration caused the gyro to saturate and drove the counter to its maximum value. This input to the ACE caused the 12 second fire of thruster 4. The burn would probably have been even longer than that except that the automatic abort circuitry shut everything down -- but the damage had already been done.

Steve Agid, Spacecraft Analyst, had the Satcom - 1R Recovery Team (with Recovery upside down) hats made and distributed to everyone before the July 15th maneuver.   John Christopher didn't appreciate the humor and ordered everyone to take the hats off! - *Steve Agid Photo*

On Thursday, technicians arrived at Vernon Valley from Astro to install a "black box" in the TT&C system.   The box had a number of lights on the front panel that were supposed to warn us if we were in danger of another erratic thruster firing.   The box took Yaw Error data in the telemetry and converted it to Yaw Rate (there was no yaw rate in telemetry).   If the Yaw Rate saturated, the RED light came on.   The criteria was that if Yaw Rate saturated for 3-seconds, abort the maneuver.

On Friday July 15 we scheduled an F-1R North/South maneuver using the Astro Black Box to warn us when to abort the maneuver.   There was a full house at Vernon Valley for the maneuver.   Steve Fox, Warren Manger and a couple more engineers from Astro were on hand. John Christopher, Americom VP Technical Operations and Joe Schwarze, Director Space Systems were also there in addition to Joe Elko and his engineers.   The maneuver was scheduled in the late afternoon.   We had a number of technical discussions with all of the Astro and Americom personnel to agree on the procedures to be used and particularly those for maneuver abort.   We went through all the procedures with the Spacecraft Controllers.   Duane McMillen, Vernon Valley TT&C Manager would man the Command Generator to send abort commands manually if necessary.

We agreed that if the maneuver started to go bad, I would make the abort decision.   I was standing directly in front of the black box with Steve Fox on my right side, Joe Elko on my left and John Christopher looking over my shoulder -- no pressure at all!   The maneuver, although nerve wracking, actually went well.   Several times during the maneuver the black box red light came on for a second or two and I was ready to call ABORT but then the light went out.   We completed the maneuver as planned and got 49 minutes and 58 seconds of burn. The Astro engineers thought that they had a good solution and recommended that the black box and its warning lights be installed in our TT&C consoles. Even though the maneuver was successful, I did not look forward to doing 10-years of maneuvers like that -- teetering on the edge of catastrophe every few weeks.

During all of this, Bob Youngblood had been heavily involved in helping to figuring out what had gone wrong.   More importantly, he had been doing calculations of the attitude disturbance torques caused by the EHT thruster plumes and what level control torques were required from thrusters 1, 2, 3 and 4 to maintain spacecraft attitude.   It was the plumes from thrusters 1 - 4 firing in Coarse control mode that was exciting the array at its resonant frequency and causing us such trouble.   After considerable work, he determined that the control firings in Coarse mode provided substantially more torque than was required to maintain attitude control while the EHT's were firing.   He then examined the control torques provided by thrusters 1 - 4 when operating in Fine mode. Fine mode used very small pulses to "clean up" any remaining attitude disturbances after the EHT's were turned off. Bob's calculations showed that Fine could provide more than enough torque to control attitude during the EHT burn.   Bob's revelation was at first greeted with disbelief but as others ran the numbers they agreed that Bob was right.

We scheduled the next F-1R North/South to be done in Fine Mode. Joe Elko, Bob Youngblood and other Americom engineers were on hand in case we got in trouble. I was positioned in front of the Astro black box to watch the lights and call abort if the Red light came on for 3-seconds. Actually I was more worried that Fine mode wouldn't provide enough control and that we would have larger than normal roll and yaw errors during the maneuver. We started the North/South in Fine mode with EHT thrusters 14 and 16. It worked perfectly. Roll and Yaw were controlled well within normal limits and the black box red light never came on. We completed the maneuver without incident. That procedure was used successfully through out F-1R's 9-1/2 year lifetime. F-2R launched in September 1983 had the same North/South configuration as F-1R and the same N/S Fine mode was used with it.

I have talked quite a bit about firing thrusters and that of course requires fuel -- for Satcom the fuel was hydrazine (N2H4). The thrusters contained a catalyst bed of alumina pellets coated with iridium -- when the hydrazine hits the catalyst, there is a spontaneous decomposition resulting in an 1800 degree F gas output. We encountered a serious problem with the fuel systems as the spacecraft approached the end of life. The Satcom spacecraft had four, sixteen-inch. spherical fuel tanks -- two on the east side and two on the west side.. One pair of tanks, one tank on the east side and the other on the west side supplied the odd numbered thrusters. The other pair of tanks, plumbed in an identical manner supplied the even numbered thrusters.

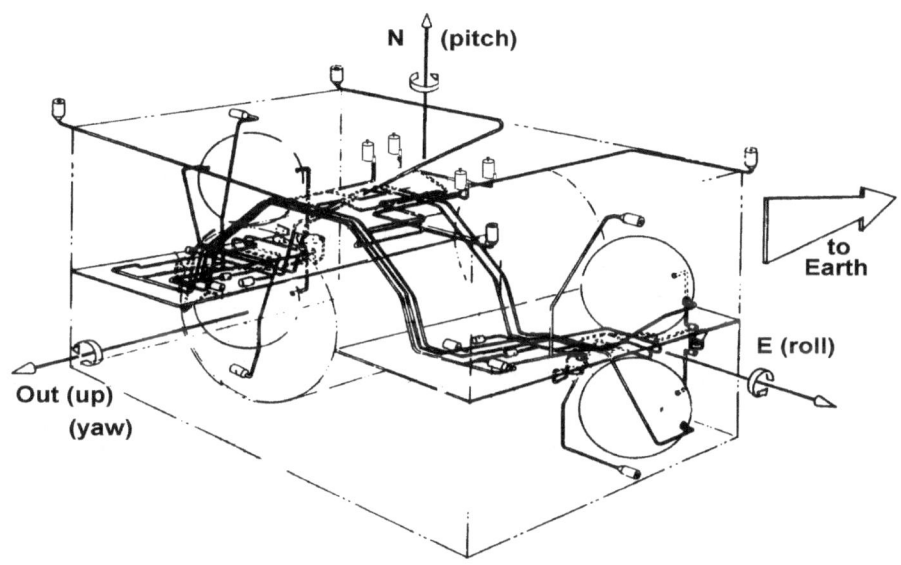

Satcom fuel system showing the four spherical tanks, thrusters and plumbing. The Northeast and Southwest tanks are for the Odd System and the Southeast and Northwest tanks are for the Even system. Thrusters 1, 2, 3 and 4 used for North South control are at the four corners of the North face. The four EHT's are in the center of the North face around the N (pitch) axis.
- *RCA Astro-Electronics Division Diagram*

One would think that the total fuel in each system (Odd & Even) would be divided equally between the two tanks in that system. That is only true if the tank temperatures are about equal. At spacecraft dawn when the sun is on the East face of the spacecraft, the Northeast and Southeast tanks get warm, the fuel and gas pressurant expand and force fuel over into the west side tanks. At spacecraft dusk when the sun is on the West face, the Northwest and Southwest tanks get warm and fuel is forced into the East tanks. Throughout the day as temperatures change fuel is pumped back and forth between the two tanks in each system. Over spacecraft life, fuel is consumed by spacecraft maneuvers and ultimately the tanks become close to being empty. At that point, the daily thermal pumping could cause the warm side tank to become completely empty at times. In earthbound tanks, this could be a problem -- in the microgravity of space it is a much bigger problem.

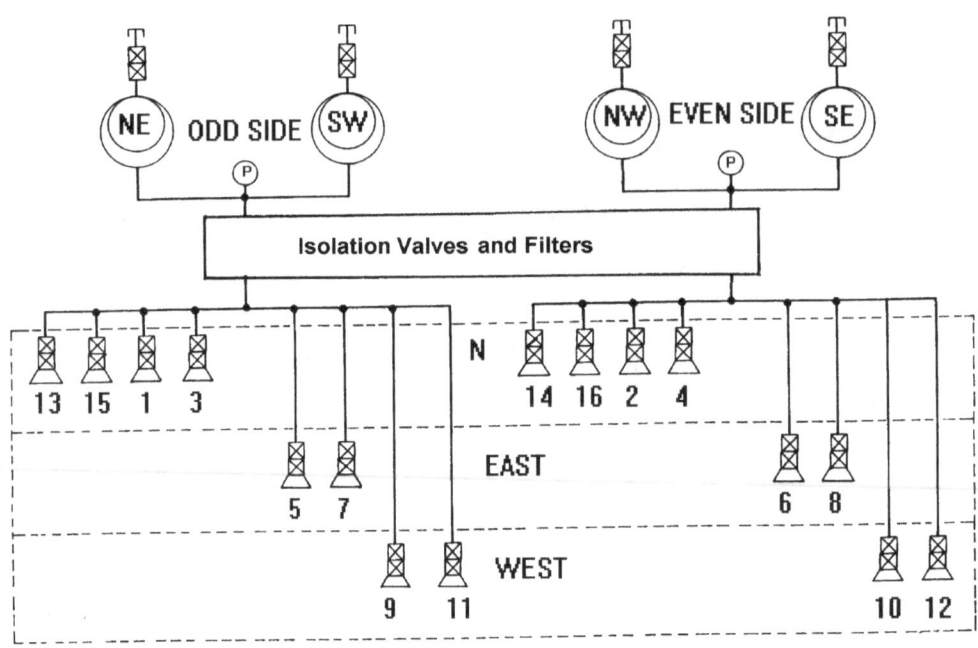

A schematic diagram of the fuel system and thrusters.   The T units on top of each tank represent the valves and fitting used to fill the tanks with fuel.   The curved lines within the tank circle represent the Fuel Management Devices or "wicks" that collect fuel to the tank outlet. The P in a circle represents the fuel system Pressure Guages - *RCA Astro-Electronics Division Diagram*

In the space environment the fuel "floats" around in the tank. As it does so it encounters the the metal plates of the Fuel Management Device (FMD) and viscosity and surface tension causes the fuel to run along the PMD into the perforated sump for the outlet pipe.   If the sump becomes dry and thrusters are fired helium pressurant will go to the thrusters not fuel. We understood this but thought that it was not a problem as long as the tanks were at least 5% full. We were prepared to use tank heaters to try to balance temperatures and prevent pumping as fuel got low.

Our Alascom spacecraft, Aurora 1, also called F-5, was nearing the end of life.   The 296-pounds of fuel loaded on board before launch were now, by our bookkeeping, down to about 23-pounds or about 8%.   We thought that we had about four or five months to go before fuel pumping was a problem. In January 1991, we were doing a 40-minute North/South maneuver on F-5. Remember, F-5 did not have EHT's. During the last twenty seconds of the maneuver first one and then both Odd North thrusters blew helium instead of fuel. The maneuver aborted because of high attitude disturbances caused by the uneven and intermittent thruster fire. After the maneuver aborted, the attitude disturbances were quickly controlled using even thruster firing. After review of what happened, procedures were implemented for controlling tank temperatures the day before any maneuvers on F-5. Analysis of the aborted maneuver showed that about six-ponds of fuel were moving back and forth every day. At the start of the North/South maneuver, there was probably only about two-pounds of fuel in the West (warmer) Odd tank and that fuel was continuing to flow to the East side.

Before the next North/South maneuver in February, we were able to use the tank heaters to keep the East and West sides within a couple of degrees of one another. As soon as the maneuver began, we started blowing helium through thrusters and the maneuver aborted. We thought that by keeping control of temperatures and doing short firings of thruster we could clear the helium out of the fuel lines and get the fuel tank sumps rewetted. We tried various procedures to access the fuel over many days without success. Every time we fired thrusters we kept blowing helium through the them and in late March 1991 F-5 was declared lost and the Alaskan traffic was transferred to the orbital backup C-1.

Disposing of F-5 responsibly was a big problem -- we needed to prevent it from causing trouble for all of the other birds at synchronous altitude. Trying to push it from its current location at 143 West Longitude up the "gravity hill" and over the summit at 162 West was quickly deemed too dangerous. If we started F-5 westward and then ran out of fuel (or helium) before we got it "over the hill", F-5 would zing back and forth through the whole domestic orbital arc for about ten-years. F-5 was already drifting eastward down the "gravity hill" toward the bottom that was at 105 Degrees West Longitude (due south of Denver). We turned off all of the transponders and communication receivers and coordinated flybys and possible beacon interference with the other spacecraft that F-5 passed on its way to 105-West. As we anticipated might happen, fuel had returned to the tank sumps during F-5's eight month drift down to 105. When we performed the East/West stopping maneuver, the thrusters fired which was a bonus. Bob Youngblood[7] had calculated that we could get sufficient thrust to stop F-5 by just blowing helium. We spent about another three months doing orbit determinations and adjusting F-5's drift to get it stabilized in the gravity "null point" at 105 West.

7 Bob Youngblood did an incredible job of working the initial problem and then doing an extensive in-depth study and analysis of fuel systems and how they perform in microgravity. See: "Trouble With Tanks" AIAA 96-1147 Robert F. Youngblood, Senior Member American Institute of Aeronautics and Astronautics.

The author explaining the use of strip chart recorders in Spacecraft Operations to Bob Frederick, President of RCA Corporation and John Christopher, VP Technical Operations RCA Americom at the Vernon Valley Earth Station. - *RCA Americom Photo*

There are over 10,000 AM and FM radio stations in the United States. Many of these are locally owned and operated but most of them are associated with radio networks that distribute news, sports, talk and music programming nationwide. This network programming was originally carried to the local radio stations on high quality telephone lines. In 1979 ABC Radio initiated a requirement for a high quality digital compressed data distribution system by communication satellite. Americom partnered with Scientific Atlanta to develop the design and implementation of the ground system that included the gateway facility in Vernon Valley and terminals for each of the radio stations. The system took about four years to develop and used digital modulation and companding techniques to deliver high quality 15-kilohertz stereo radio programming. The programming originated from the radio network's Manhattan Studios that were connected by landlines to the Americom microwave facility at 555 West 57th Street. The digital programming was transmitted via an NEC Digital Microwave system to Vernon Valley and thence by satellite to the affiliate radio stations all over the US. Eventually, all of the radio networks including NBC, CBS, RKO and Westwood One and adopted the system. The Digital Audio Transmission System or DATS ultimately filled six transponders in F-1R and distributed programming to 4500 network affiliates with dedicated earth stations.

The DATS traffic was transmitted to F-1R, which was located at 139 degrees West Longitude, from Vernon Valley. The elevation look angle from Vernon to F-1R was about 12 degrees. Because of the low look angle, the DATS carrier was about over Wilkesbarre, Pennsylvania before its altitude was above 60,000 feet and out of the reach of thunderstorms. All of the networks and their affiliates understood that heavy rain could interfere with the DATS uplink and they could cope with it -- as long as they were notified in advance. My problem was that the sun could very well be shining in Vernon and there could be a thunderstorm cell interfering with DATS in eastern Pennsylvania. I entered into a contract with AccuWeather in College Station, PA to warn us when heavy rain or snow could be expected in the DATS path. I provided AccuWeather with a geodetic chart showing the DATS beam path until it reached 60,000 feet. AccuWeather sent warnings to the Americom NOC with expected times of heavy precipitation passage through our beam. The NOC would immediately pass that info to all of the networks in New York and they would sent messages through the DATS network to printers in the affiliate stations. It all worked very well. Actually DATS was a fairly robust system and it took pretty heavy rain to take it off the air.

Chapter 14 - Baby Killers

Satellite communications share the same microwave frequencies used by terrestrial microwave communications systems. To prevent interference, satellite earth stations are generally located in rural valleys where the surrounding mountains shield them from microwave radio signals. Vernon, New Jersey is located in a valley 600 feet above sea level surrounded by 1400 foot mountains. When RCA first wanted to build a satellite earth station in Vernon Township, New Jersey, the town fathers and mothers welcomed them with open arms. The township already had two earth station facilities, American Satellite Corp. and Western Union Communications. The township considered the earth stations to be excellent ratables for the township tax base. The stations paid substantial taxes but did not put many children in the school system. Vernon told RCA that for land use purposes the township would treat earth stations as utilities and thus they would be a Conditional Use in any zone in the town. The RCA Americom Earth Station Facility ended up in the 136 acre former Sam Edsall farm that was zoned residential. In order to build facilities or add on to them RCA had only to present a proposed Site Plan to the Planning Board for approval. Since the first construction in 1975 and up to 1982, the system had worked well and RCA had added 4 C-Band antennas, expanded the building and added a parking lot. with no problems.

The Vernon Valley Earth Station in 1982. An extension had been added to the building, the lower parking lot had been built and four satellite antennas had been added. Note the 11-Meter antenna on top of the tall tower so that it could "see" F-5, the Alascom bird, out over the Pacific Ocean at 148 degrees West Longitude. - *RCA Americom Photo*

The earth stations also required microwave towers on the earth stations themselves and on the surrounding Waywayanda and Hamburg Mountains to relay their communications traffic back and forth to New York City. Eastern Microwave Inc. also operated a microwave facility on Hamburg Mountain to distribute television programming from New York to cable systems in New Jersey and Pennsylvania.

At about this time, however, some citizens groups in various parts of the US were raising concerns about the possible health effects of exposure to microwave radio signals. Americom's application for an earth station on Bainbridge Island to serve the Seattle, Washington area had been opposed by local activists. Because of this, RCA was looking for another Puget Sound location that would be less susceptible to citizens objections. A book by Paul Brodeur, "The Zapping of America" was getting a lot of attention by the activists. Unfortunately, Brodeur's book was factually incorrect about the effects of non-ionizing radiation and contained misleading information about the incidence of "medical clusters" supposedly caused by these effects.

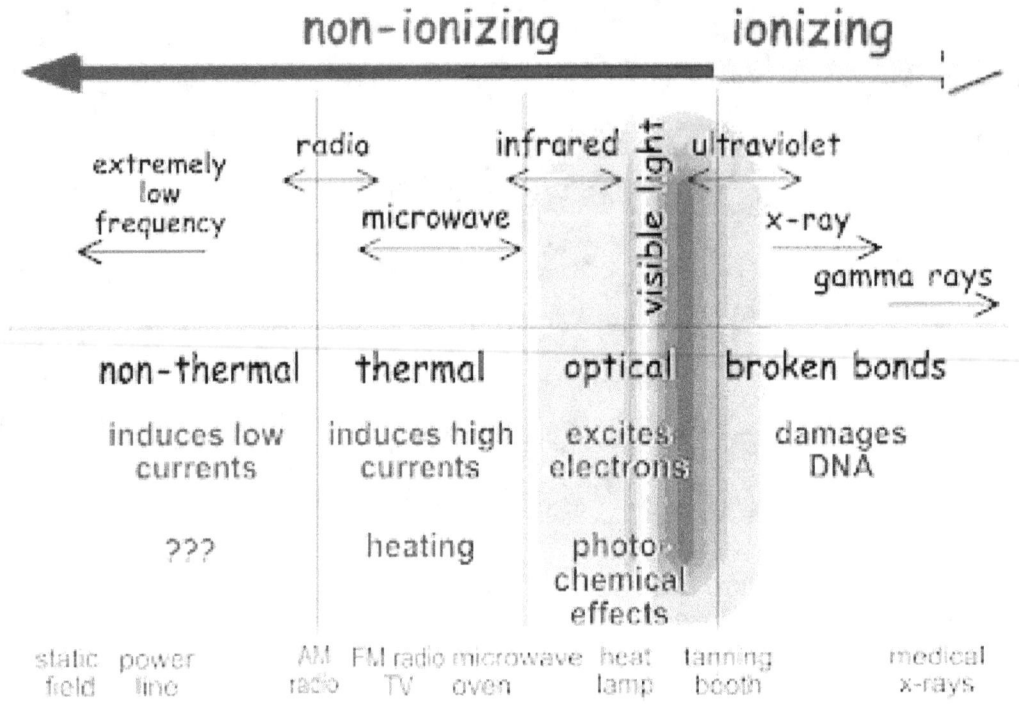

The Electromagnetic Radiation Spectrum showing the ranges of non-ionizing and ionizing radiation - *Environmental Protection Agency*

The medical effects of microwave radio signals have been studied extensively since they first came into widespread use during World War II. It is important to understand that microwave radio signals are classified as "non-ionizing" radiation in contrast to X-Rays and Gamma Rays that are classified as "ionizing" radiation. Electromagnetic Radiation covers a very wide spectrum as can be seen by the chart above. The lowest frequency waves in the spectrum include electrical power at 50 or 60 Hertz and extend upward through radio and TV, microwaves, Infrared, Visible Light, Ultraviolet, X-Rays and finally Gamma Rays. The frequencies below ultraviolet are non-ionizing and those above ultraviolet are ionizing. Ionizing radiation such as X-Rays and Gamma Rays carry enough energy to completely remove an electron from an atom or molecule thus damaging that atom or molecule and possibly causing cancer in human beings.

186

Nonionizing radiation adds energy to the electrons through excitation which in the case of microwaves can cause heating but not damage, to the atom or molecule. The most common application of this phenomena is the Microwave Oven that uses a microwave power source at a frequency 2.45 GHz to cook food by dielectric heating. The microwaves primarily effect water and other liquids ("add a teaspoon of water before cooking").

The manufacturers and users of microwave equipment such as radars and microwave communication equipment had long recognized that at high power levels microwaves could effect the human body. Study had showed that soft tissue (eyes, testicles) were susceptible to heating damage. The American National Standards Institute (ANSI), working in conjunction with industry, military, academic and scientific organizations had established a maximum allowable human exposure level of 5000-microwatts per Square Centimeter in the frequency range of microwave communications. Any levels below that were considered safe. All of us who were involved with earth station and microwave communications were not concerned because we knew that, at most, microwave radio signals in our facilities were no more than 5 or 10 microwatts/cm^2 -- thousands of times less than the standard

RCA Americom's business was expanding and more satellites were required. In the fall of 1982 Americom made an application to Vernon Township to add another C-Band antenna and to relocate the existing microwave tower closer to the equipment in the building. No one was concerned about the township approval process. As was usual, when the Americom application was ready for submittal to the township, we had Harald Harvey our Facility Technician hand carry it over to Vernon Town Hall. Harald was active in the Sussex County Republican Party and was on a first name basis with all the local politicians -- couldn't hurt! Harald came back with some disturbing news. His contacts in Town Hall told him about a new organization called Vernon Citizens Against the Towers (VCAT) that planned to oppose expansion of any of the earth stations in town. The leaders of VCAT were Steven and Elise Kreindler who lived in Mayflower Estates, the subdivision immediately up the hill from and abutting the Americom property. We passed on as much details as we had to Carl Cangelosi, Americom VP and General Counsel and David Connelly, the attorney that would represent Americom at any land use hearings in Vernon.

Americom did some more digging about VCAT and found that they were going to hold a press conference at the Kreindlers home. They were media savvy enough to get the conference listed on the UPI Day Wire so they were assured of good attendance.   Carl Cangelosi called me and said that he would like me to be the official spokesperson for Americom on the microwave issue.   I quickly reviewed everything that I could find about microwaves and health and practiced some "sound bites" that summarized our position.   "These stations are not hazardous, microwave levels are thousands of times lower than the ANSI Standard."   "We have nothing to hide -- Vernon Township can select an engineering firm to make measurements and RCA will pay for it."

I didn't have to wait long.   Christine, my secretary, called in to me "there is a TV crew at the front door".   Actually there were a couple of crews.   The VCAT had their press conference at the Kreindlers home about a quarter mile up the hill from the station.   They had about half a dozen mothers with children having Down Syndrome or other disabling birth defects at the conference.   Elise Kreindler opened the press conference stating that Vernon Township had higher number of birth defects than other parts of Sussex County and they were all caused by microwave radiation from the earth stations.   Each of the mothers holding their children told the story of how they believed that the satellite stations were responsible for their personal tragedies.   As soon as the press conference was over, the media came down the hill and arrived in my lobby.   By and large they were fair and wanted to get both sides of the story.   That night the interviews were on the New York TV local news shows.   I thought they went reasonably well.   Carl called me the next day and congratulated me -- he thought it went as well as could be expected when we were being accused of deforming babies.   VCAT knew that we had an application for an antenna and tower relocation filed with the township and they publicly vowed to fight it with all the forces that they could muster

Shortly thereafter the other shoe dropped. Emmanuel "Manny" Honig, dean of the Sussex County Bar and attorney for the Vernon Township Zoning Board wrote a letter to the Township Committee stating that the township had erred in treating satellite facilities as Public Utilities. As a result of that ruling, RCA Americom's facility now became a "Pre-Existing Non-Conforming Use" in a residential zone. It also meant that RCA could no longer just submit a Site Plan to the Planning Board for approval. Now we would have to make application to the Zoning Board for a zoning variance for any site expansion. That would undoubtedly be considerably more difficult than our previous Planning Board experience and be further complicated by the VCAT opposition.

There was considerable high level discussion within Americom of how to proceed. What if Americom's Vernon facilities were frozen in their current configuration? How would the TT&C be able to support new spacecraft? Americom was also embroiled in a similar microwave radiation dispute while trying to build an earth station to serve Seattle. It was a frustrating situation because we knew that there was little or no microwave radiation outside the boundaries of our facilities. In addition we also knew that there was no scientific evidence linking microwaves and birth defects. However, we also knew that children's health was a very sensitive issue and we had to be extremely careful in discussing it in public. Americom hired Dr. Bill Guy from the University of Washington, an internationally recognized authority on the effects of microwave radiation, to advise us and to be an expert witness at any public hearings. John Williamson, Director of Public Affairs, Carl Cangelosi and I would be Americom's public face on all microwave related issues.

The VCAT people were very active. They did a "survey" of birth defects in Vernon Township and put together a map of the town with colored pins denoting the location of different types of defects.   The data for the VCAT "survey" was evidently gathered by word of mouth and telephone calls -- no official records were consulted. VCAT also put out a press release listing "clusters" of birth defects that they had discovered during their polling.   They said that there were clusters in a number of lake communities in the township.   Going through their press release it didn't seem to me that any of these places were anywhere near our station and that there was a lot of terrain intervening between us and them. I took the VCAT "data" and topographic maps of Vernon Townships and made vertical elevation plots along a line of sight from the Americom station and the supposed cluster.   In all cases there was a hill or mountain obstructing the path -- in other words a beam from our antenna would have to go through solid earth and granite to illuminate the cluster area.   I forwarded the plots to John Williamson and he had them drawn up and slides made.

Every time VCAT did something or made some other claim, they issued press releases or held press conferences.   As a result, I was spending a lot of time on the phone with reporters and standing in front of TV cameras.   The local New York TV reporters were by and large fair in their coverage.   The crews consisted of a camera person, an audio person and the on-air "talent". They came to the station, talked to me while I gave them a tour and then we did an interview in front of one of the antennas.   Then the "talent" did the on camera "stand up" monlogue and they were on their way.   They just wanted to get the story straight, do what they had to do and get the tape back to Manhattan in time for the "Six O'Clock News".   One time I was a bit concerned about the coverage. Channel 9 called one afternoon and said that an INN crew would be at the station shortly.   The sun was going down by the time that they arrived and they immediately started shooting Sheila Stainback's stand-up in front of the antennas -- they wanted it done before the sun went down.   I was very concerned because I had not had an opportunity to present Americom's position and her comments were already "in the can" before I talked to her. I needn't have worried. Ms Stainback had obviously done her homework and did a very balanced reporting job.

It wasn't as easy with the national network folks. In addition to the crew noted above they also brought along a producer who decided how the story would be shot and how it would be reported. I spent most of a whole day with a CNN crew.   As we discussed the microwave issue, Scott Berrett, the "talent" seemed to agree with our position on microwave safety.   His producer, however, was very skeptical and kept bringing up VCAT talking points.   We talked a lot both off and on camera and I guess I partially convinced the producer. The CNN piece, was about 10-minutes long and turned out to be reasonably well balanced.   CNN would return on two more occasions before this was all over.

The crew from "The CBS Evening News with Bob Schieffer" was a lot tougher.   They were with us for two days and the producer was bound and determined to get an incriminating sound bite out of me.   They set up my interview in front of the antennas and they even had a tripod for the camera (no hand-helds for the network).   The producer grilled me for about a half hour. She started with a couple of fairly innocuous questions and then asked "Isn't it possible that these antennas might pose a hazard to the community?"   I answered "Absolutely not -- the signals from these antennas are thousands of times less than the ANSI standard". She asked a couple of more questions and then came back to the hazard question again.   My first thought was to answer it differently, but then caught myself and gave the same answer.   She kept it up throwing in the hazard question about every three or four questions and I gave the same answer. I am sure that if I had ever answered "Well it might be" that bite would have been the lead on the CBS Evening News.   I thought that we were finished with CBS when on the morning of the third day, I saw that the crew was set up down by Route 517 where they had a broad view of all of our antennas.   I'm thinking "What the Hell are they up to now?" At this point a big white limo glides up and out steps Chris Kelly the "talent" for this piece.   The producer hands him his microphone, he does the stand-up, gets back in the limo and away they all go. The piece, when it aired, was actually pretty balanced.

One day in mid December, Carl Cangelosi called me and said that VCAT had arranged to be on a talk show called "Make Peace With Nature".   Could I go with him to WKRC in Cincinnati the next day to tape two half-hour shows.   (It really is WKRC not WKRP) I said "sure -- what was the deal".   He said that it was an nature program underwritten by the Izak Walton League that examined environmental issues.   Elise Kreindler and I would be on the program with host David Surber discussing the Vernon Microwave controversy.   Carl would go along to be sure that we were treated fairly and to protect RCA's legal interests.

We arrived at WKRC and Surber's office at about 7PM.   Elise was already there and had a foot-high pile of books with her -- I never figured out what they were for. Elise Kreindler was an attractive Danish woman who had attended the Royal Theater Drama School.   I always suspected that Elise had latched on to her spokesperson role as a step to a television career. She had no personal interest in the issue -- the Kreindlers had one 14 year old perfectly normal daughter.   We had a brief discussion with Surber and his producer and floor manager about how the talk format would work and then went out to the set.   Actually Surber was very fair and tried to give each of us equal time.   I had reviewed all of VCATS press releases and Elise's interviews so I had answers to her allegations.   In the second half hour I introduced the mountain profile slides and explained the impossibility of the VCAT theory. Elise kept repeating her opinion that it didn't matter -- microwaves could go through mountains.   Carl and I both felt that it had gone well -- when we saw the final broadcasts, we felt even better.

As December 1982 came to an end, it was apparent that this was going to be a long slog.   The township had selected Hamilton Communication Consultants to do the impartial measurements of microwave radiation in the township and that could take many months.   The Sussex County Board of Health and the New Jersey Department of Health had been contacted by VCAT and they said that they were looking into the situation.   VCAT had also contacted the Centers For Disease Control in Atlanta and the U.S. Environmental Protection Agency (EPA) with their claims.   Americom management decided to withdraw the pending application for another antenna and the tower relocation.   The thinking was that if we were going to go through all of this, we should ask for everything we might need for the foreseeable future and get it over with.   The Americom Marketing and Engineering Departments went to work to develop a plan for the Vernon Valley facilities expansion required to support long term market growth.

As the plan progressed, great care was taken to consider how each facet of the proposal would be presented to the Vernon Township Zoning Board of Adjustment..   The current land use regulations did not require an Environmental Impact Study, however, Americom retained an environmental consulting firm to prepare one.   It became apparent that the plan would require an additional building.   An architecture/engineering firm owned by a Vernon Township engineer with offices in Vernon was selected to design the new building.   A landscape architect was retained to design landscaping ($280,000 worth) to shield the Americom facilities from view.   A model of the station showing how it would look when the expansion was completed (including landscaping) was built to be used at the hearings.

The attacks by VCAT were relentless.   Every week there was at least one article in the local papers about how RCA was causing deformed children in Vernon Township.   Elise would find various pretenses to get sound bites on WSUS the local FM Radio station.   It got to the point where our employees would not wear their RCA jackets in town.   My kids reported overhearing comments about "baby killers" at school.   And if we ever suspected a conspiracy, we got, what many of us felt was absolute confirmation.   Manny Honig, the Vernon Zoning Board attorney who had written the letter ruling against RCA, resigned from the Zoning Board and became the attorney for VCAT.   So now our nemesis was going to lead the fight against us when our application was heard in Vernon -- have a nice day!

Through all this, I was still answering reporters questions, appearing on TV and speaking before Rotary and other groups defending our position.   We were also working on our image in town -- when the Board of Education came up short for a playground at a new school, RCA donated the money.   Although, I believed all of the technical data, I still wanted to assure myself that what I was saying was absolutely so.   I asked Jim Evans, the earth station manager, if we could make some measurements ourselves just to be sure. Jim had a 6GHz (our transmit frequency) test horn connected with low loss Gore cable to an HP Spectrum Analyzer to measure the levels.   We took that equipment and a small portable generator to power the test equipment in the back of his pick up truck.   First I went right out in front of the 13 Meter Vertex Antenna that was transmitting a number of saturated carriers to F1R.   Because of the low look angle the edge of the dish was only about four feet off the ground.   I had the test horn on a pole and at about 15-feet in front of the dish with the horn four feet above my head we measured about 200 uWatts -- remember the ANSI standard is 5000.   We measured all around our property edge and if we were able to find any signal at all, they were mostly about 0.01 uWatt.   The strongest signal we could find at the property edge was at the top of our sloping driveway behind the antennas where we measured 0.02 uWatt.   That was probably back splatter from the antenna subreflectors. I definitely felt much better.   We also went into a couple of the lake communities where the VCAT "clusters" were reported -- we could find no measurable microwave signals. We went to the top of the ridge line on Lake Walkill Road where we had a direct view of the massive AT&T Microwave Facility about 10 miles away near High Point.   There were at least 10 Microwave Horns transmitting toward us -- we could find no measureable signals. Now I could speak with first hand knowledge that there were no damaging microwaves emanating from the RCA Vernon Valley Earth Station or anywhere else near us.

As we got into 1983, Hamilton completed its microwave measurements and delivered its report to Vernon Township.   These were highly accurate measurements with test equipment whose calibration was traceable to the US Bureau of Standards. That report stated that the strongest signal that they found at the RCA property line was 0.01uWatt.   They also reported that the strongest signal they found at American Satellites border fence was 0.13uWatt. Months later when the US EPA brought in their Mobile Monitoring Unit from Las Vegas, they obtained readings similar to those measured by Hamilton. The NJ Bureau of Radiation Protection did a survey with a Narda Meter designed to detect signals exceeding the standard of 5000 uWatts so of course they detected nothing.

Throughout 1983 we continued working on the Vernon Valley Plan. The objective was that, if approved, we would not have to go back to Vernon for any more improvements for many years. Comprehensive site and building plans were drawn and bids were obtained for all of the construction and improvements. The environmental consultant did a very comprehensive study of the 136 acre RCA property and completed the Environmental Impact Statement. The Landscape Design was completed including an irrigation system with its own deep well to supply water. The VCAT drumbeat had slowed down but not stopped. The various microwave radiation tests were complete and all of them showed insignificant microwave levels in Vernon Township. RCA did its best to insure that these results were widely reported in the local media. Preliminary results from county and state health departments review indicated that Vernon birth defects appeared to be within normal levels. According to Margaret G. Conomos, a research scientist in the environmental health program of the state's Division of Epidemiology and Disease Control, the state's study of Vernon showed that, from 1975 to 1981, there were 57 birth defects in 1,653 pregnancies in Vernon Township and 333 defects in 10,351 pregnancies in Sussex County. The study showed an annual birth-defect rate of 31.9 per 1,000 pregnancies in Sussex County, compared with the annual national average of 33.9 per 1,000 pregnancies. Miss Conomos said that health officials did not find those statistics significant. The US Center for Disease Control also reported that the birth defects in Vernon were within the normal bounds. Ultimately, a couple of years later, the NJ Department of Health Report on Vernon stated that the VCAT survey "had no statistical value".

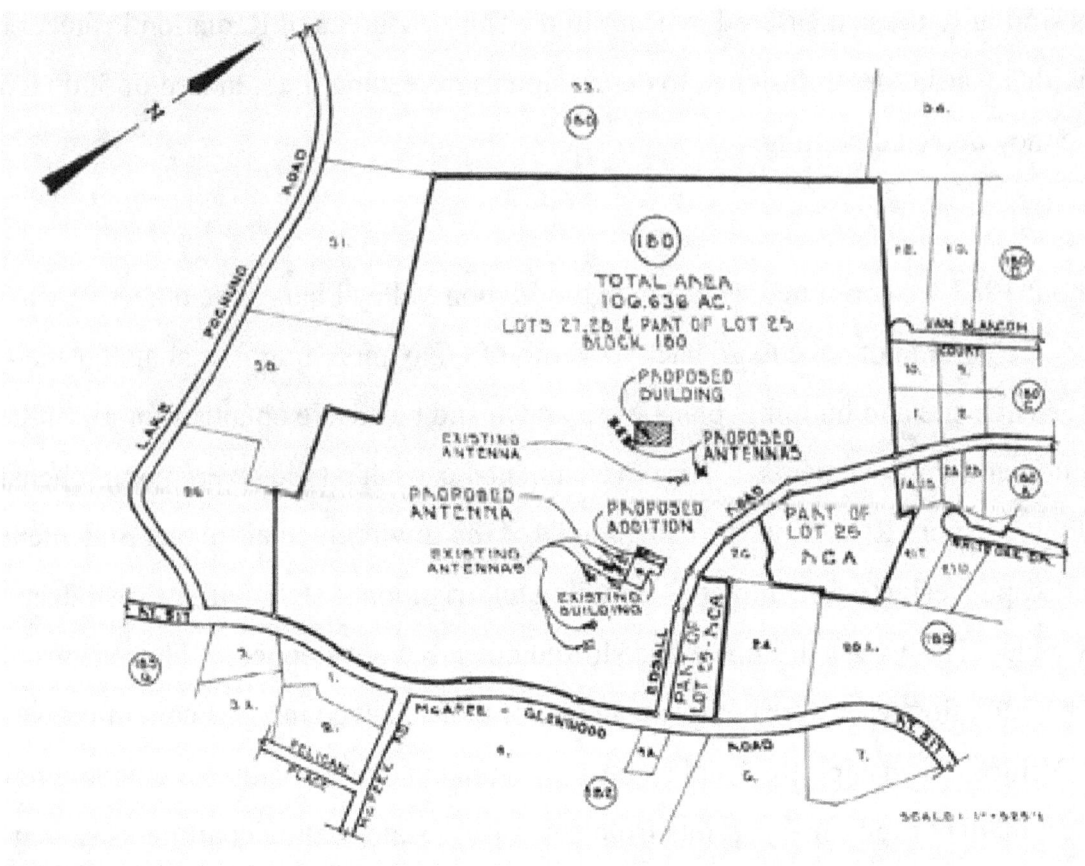

In the February 1984, RCA Americom submitted their $12.5 Million Vernon Valley expansion plan to the Vernon Township Zoning Board of Adjustment (see plan above). Besides all of the land use data, it also included the Environmental Impact Study and the Landscaping Plan. The model of the site was delivered to the Township Hall for the hearings. The hearings got started and were attended by huge crowds. In fact the crowds were so large that hearings were initially held in school auditorium rather than the town hall meeting room. As the hearings dragged on, the crowds got smaller and we ultimately ended up back in the town hall. Altogether the hearings lasted nine months. It took us a number of meetings to present our case. Dave Connelly, our attorney, and I presented the land use and technical plan. Our Environmentalist presented the Environmental Impact Statement. Our Arborist presented the Site Preparation and Landscape Plan. Dr. Bill Guy gave an extensive presentation on the electromagnetic spectrum, the characteristics of and techniques for measuring microwaves. A Medical Doctor from the Miami VA Hospital presented on the medical effects of microwaves.

Finally after nine months of meetings, and testimony, the board voted and approved our application.   Anticipating that VCAT might appeal the decision, RCA was poised and ready. Construction started within a few days and progressed rapidly on the new building, antennas and landscaping.

VCAT ultimately filed suit in New Jersey Superior Court in Morristown, NJ.   By the time it was finally heard by Judge Reginald Stanton, all of the construction approved by the board was complete.   Judge Stanton ultimately ruled against RCA and imposed a fine and a 15-year facility expansion moratorium.   I reviewed the transcript of the hearing and saw that Judge Stanton had made several serious technical errors that could very well reverse the decision. At that point, however, RCA Corporation wanted to put this suit and another with Showtime behind them so they decided to settle both. I was disappointed because I felt that we could have won in court and now we had to endure a 15 year hiatus in construction -- it was fortunate that we got as much done as we did.

The good news, if there was any out of this mess, was that as a result of the RCA Americom and Vernon Township debacle, the State of New Jersey changed their land use regulations. Municipal zoning and planning boards are now prohibited from considering any medical issues in their deliberations.   They now have to come to their decisions based purely on land use considerations. Any medical issues can only be addressed by the State of New Jersey. Although they undoubtedly don't know it, petitioners for the literally hundreds of cell towers that have been approved and built in NJ in the past 20 years have us to thank that they didn't have to deal with "Citizens Against The Towers"

Chapter 15 - A Tough Business

From the inception of RCA's satellite communication enterprise, telephone communications had been considered central to the business. The primary driver for the RCA proposal for a four-satellite domestic system in 1972 was to provide telephone service to Alaska and to connect the Alascom telephone system to the lower 48. Before RCA even had a satellite of its own, RCA Globcom built earth stations in Point Reyes (San Francisco) and Valley Forge (New York) to carry telephone message traffic to and from Alaska. The Private Leased Channel (PLC) business was considered central in the formation of Americom. At the outset, earth stations were built in Vernon (New York), Lake Geneva (Chicago), Rayburn (Houston) and South Mountain (Los Angeles) in addition to the existing Valley Forge and Point Reyes stations. These locations served cities where it was expected that there would be substantial PLC business -- and there was.

PLC was not an easy business from a technical standpoint. Customer telephone circuits arriving in the Central Telephone Office (CTO) were bundled together in Groups (12 telephone circuits) and Super Groups (5 Groups or 60 Circuits). The process of doing this combining of circuits is called multiplexing and the actual technique is called Frequency Modulation/ Frequency Division Multiplexing (FM/FDM). For our purposes that is all you need to know about this fairly complex system. The multiplexed (muxed) circuits are transmitted through a microwave radio link on tall towers between the CTO and the Satellite Earth Station. In the earth station the circuits are further combined into 16 Supergroups plus one Group for a total of 972 telephone circuits, the maximum capacity (at that time) of a satellite transponder. At the far end the process is reversed.

Now comes the hard part -- getting the telephone circuit from Americom's CTO to the customers location. First of all, telephone circuits in microwave and satellite systems are handled as four wire circuits -- two wires (a pair) to transmit and a pair to receive. The telephone on your desk uses only one pair to transmit and receive. This means that either in the CTO or on the customers premises there has to be a "customer package" containing the electronic circuits to convert from a four wire interface to a two wire interface. Whether it is two-wire or four-wire the only way to connect the "last mile" between the CTO and the customer is in copper wires. Virtually the only way to do this is by leasing telephone circuits from the local telephone company (New York Telephone for instance in Manhattan). Although required by law and regulation to supply these circuits to Specialized Common Carriers (SCC) like Americom, the local telcos were in no rush to help their competitors. It usually took about 6 weeks for the telco to get a "local loop" installed.   Once installed and in use, 84% of interruptions to service were caused by problems with the local loops. Of course all of the SCC's and AT&T depended on telco supplied local loops and we all had the same reliability problem with them.   To differentiate Americom's business from its competitors required an outstanding customer support organization and circuits designed for reliability.   In many cases Americom ordered extra local loops for a customer so there were spares available in the event of telco failures.

In the early days of Americom's PLC business in the late 1970's, RCA considered our primary competitor to be Western Union (WU).   At that time AT&T was constrained by court order from entering the satellite PLC business until 1979 and American Satellite was a small startup serving only three cities.   Western Union had Westar, the first domestic communication satellite system, that was up and running more than a year before Americom's first launch. The Westar satellite system served Seattle/Portland, San Francisco, Los Angeles, Honolulu, Dallas, Chicago, New York and Huntsville/Chattanooga.   WU also had a national microwave network connecting many smaller cities and towns. WU had been a pioneer in microwave communications building a mid-Atlantic network between New York, Philadelphia, Washington and Pittsburgh in 1945 -47. In the 1960's they had expanded that microwave system into a true transcontinental network serving over 160 cities.   RCA expected that WU would market their combined satellite and microwave networks as a complete integrated communications system. Never happened. Evidently the satellite people and the microwave people instead of working together fought internally over who would serve different areas or customers.

My observation of the Westar satellite operations was that they were badly managed and suffered from a lack of in-house spacecraft engineering support.   Nothing illustrated this better than the TDRS debacle. Western Union entered into a lease service contract to provide NASA 10 years of space communications services with a Tracking and Data Relay Satellite System . The project also included what was called Advanced Westar with the first two TDRS spacecraft planned to replace Westar's 1 and 2. The TDRS spacecraft payload consisted of an S-Band system for NASA's Tracking and Data Relay requirements, a 24-transponder C-Band system for Advanced Westar and a Ku Band system to be shared by NASA and Westar. Western Union Space Communications (WUSC) was responsible for the spacecraft bus and the ground segment. WUSC subcontracted the spacecraft design and construction to TRW. One day I got a call from Jim Gregory at Westar asking me whether North/South station keeping was really necessary.   I told him of course it was -- if the bird wasn't kept in the FCC defined box, customers would require expensive tracking antennas to follow its daily motion. A little while later, I got a call from Jim Judson, the Westar Spacecraft Manager and a former colleague of mine from RCA Service Co.   Jim was also curious about how necessary North/South stationkeeping was.   As we talked Jim revealed what the problem was. Evidently the Westar folks had just discovered that the TDRS birds, being built by WUSC, that was going to carry their Advanced Westar communications did not include fuel for North/South stationkeeping.   Without North/South stationkeeping TDRS would be useless for commercial communications.   Evidently when the system designers calculated the weight for launch on an Atlas launch vehicle, they didn't include the 1000 pounds of fuel for North/South stationkeeping. Obviously there was no one in WUSC that truly understood commercial satellite communications.   Westar told WUSC, TRW and NASA that Advanced Westar would not work without North/South stationkeeping.   This caused the TDRS/Advanced Westar program to suffer an enormous setback in schedule and costs. This requirement meant that they had to go from an Atlas expendable launch vehicle to a Space Shuttle launch in order to carry the additional fuel weight. It really didn't matter because Westar would be out of business before an operational TDRS got launched. Without TDRS and its Advanced Westar, Western Union had to launch Westar-4 on February 26, 1982 and Westar-5 on June 9, 1982. After suffering heavy losses, Western Union sold Westar to Hughes Communications in 1988. TDRS-1 failed to achieve synchronous orbit and TDRS-2 was destroyed in the Challenger disaster. TDRS-3 launched from the Space Shuttle on September 29, 1988 included the Advanced Westar transponders, however, the antenna pattern had been adjusted to allow their use in international communications. In mid-1989 NASA requested proposals for a six year lease to use the C-Band transponders. Columbia Communications, a Honolulu based company won the rights to use the C-Band transponders in the Pacific and Atlantic TDRS birds for international traffic.

PLC was a tough competitive business. As noted before there was a lot of competition from other Specialized Common Carriers (SCC) such as AMSAT, Westar, Sprint and MCI. In addition, after 1979, AT&T was getting more aggressive in the face of SCC competition.. As a result there was a lot of "churn" in everybody's customer base. Most contracts were for a year and as soon as the year was up customers would leave for a better deal from another carrier. In one year, it was difficult if not impossible to recover all of the installation costs. Also there was a lot of "onesy, twosey" in the business. Large companies with many AT&T PLC circuits would give one or two circuits to each of the SCC's to prove to their shareholders that they were interested in competition and lower costs. In many cases, Americom was able to build on the one or two circuits by providing good customer service in addition to low cost communications. Americom was committed to the business and added customer service and marketing capability.

Despite the problems and complexities of the PLC business it grew and Americom expanded to accommodate it. Ultimately there were Americom CTO's in New York, Los Angeles, San Francisco, Miami, Camden, NJ, Boston, Houston, San Antonio, Dallas, Honolulu, Seattle, Kansas City, Atlanta and Chicago plus a Customer Service Center in Princeton, NJ. To carry all of this satellite traffic in the most cost effective manner, Americom Engineering and the RCA Sarnoff Research Center in 1984, developed a Single Side-Band (SSB) modulation scheme to replace the existing FM/FDM system. Using FM/FDM a single transponder could carry 972 voice grade circuits. Using the new SSB system a single transponder could carry more than 3000 voice grade circuits! This innovation not only made better use of transponder space but also reduced the costs per phone circuit. At one point, there were more than 13,000 telephone circuits being carried through the Vernon Valley Earth Station.

As the PLC business grew, we even did business with our competitors. We provided what are called inter-machine trunks to MCI and Sprint. These are circuits at the Super Group (60 circuit) level that interconnect a switching system in one city with a switching system in another city. Because we handed off the traffic to MCI at Super Group level, these did not involve any local loops and their inherent problems. My recollection is that our price to MCI was the equivalent of $215 a circuit per month. A good deal for both of us.

On January 1 1984 AT&T's local operations were split into seven Regional Bell Operating Companies (RBOCs). The RBOCs almost immediately raised the price for the local loops that we leased from them to connect our customers premises to our Central Telephone Offices (CTO). Because of the intense competition between Americom and the other Specialized Common Carriers (SCC) such as MCI, Sprint, Westar and American Satellite there was enormous customer "churn". Customers were hopping from one carrier to the next. The cost of installing and deinstalling customers cost Americom dearly. It took at least 10 months of service before a circuit became profitable. Americom was also over extended in the PLC business and had too many CTOs. In places like Camden, NJ and Kansas City, MO there was not enough business to cover operating costs. On top of all that, many customers did not like the 1/4-second delay inherent in satellite telephone service. In addition, the new fiber optic networks, that did not have any delays, represented a severe threat to the satellite based PLC business.

In 1985, RCA Corporation ordered Americom to get out of the PLC business. They wanted it done immediately because the corporation had already written down the loss. The FCC, however, said RCA could not do that. They had to transition it more slowly so that customers could make other arrangements. The customer base was sold to MCI. Dismantling and selling all of the equipment in the Central Telephone Offices and Earth Stations and our warehouses was a monumental job. The corporation wanted it done as soon as possible so much of it was sold at fire sale prices. It was a traumatic experience -- my recollection is that Americom laid off about 600 people. It was, however, a good decision -- Americom got out before the fiber networks took hold and the satellite PLC business really crashed. Serendipitously, at the time that we suddenly had empty transponders, the cable TV business started to boom and we had TV customers for our empty transponders.

Chapter 16 Shuttling to Space

Starting in the 1950's there had been a number of proposals for a vehicle that could journey into space and then return to earth and land horizontally like an airplane. In the 1960's the US Air Force had done studies of such a vehicle including the Dyna-Soar program. NASA had also done some studies and in 1968 NASA began the project that would ultimately become the space shuttle. President Nixon approved the Space Shuttle program in 1969 and work started in earnest.

The first test flights of the shuttle orbiter were in 1977. The shuttle Enterprise was taken to a high altitude on top of a NASA Boeing 747 airplane. Initially a number of tests were run with the shuttle "flying" on the back of the 747 to characterize its aerodynamics and test its control surfaces. For the first five flights the shuttle was unmanned. For the next three flights, Enterprise was manned to test flight controls. Finally Enterprise conducted five free flights where it was released from the plane and glided down to a landing at Edwards Air Force Base. I saw all of the test flights because we were carrying the NASA television on F-2. In the first two flights the shuttle had a streamlined shroud covering the rocket nozzles in its tail -- it seemed to fly rather nicely. In the next three flights the shroud was removed from the tail and it was announced that this would increase the drag substantially. Boy did it -- when released from the 747, it dropped like a rock. I kidded about it saying that the shuttle " had a ballistic glide path". The first orbital flight of the shuttle Columbia, STS-1, was on April 12, 1981. Over the next four years the three other shuttles Challenger, Discovery and Atlantis were flight qualified and put into service.

From the beginning, NASA envisioned that the shuttles would be used to launch commercial and military satellites. In fact the designed width of the shuttle payload bay was driven by the size of the secret military KH-12 spacecraft[8]. All of the commercial satellite companies were

---

8   About $3-billion was spent at Vandenberg AF Base in California for Space Shuttle launch facilities to support

approached by NASA with proposals for launch services. RCA agreed to launch two communication spacecraft on the shuttle. RCA signed up at the beginning of the process and got an "early bird special" that guaranteed that if the launch failed, RCA would get another shuttle launch free. The second launch, however, was on the usual terms -- no "do-overs". From a business standpoint, the shuttle was a good deal. A Delta launch cost about $25 million. NASA was charging about $12 million for a shuttle launch. The shuttle had the capability to launch up to three communication spacecraft on a single mission -- and did on several missions. The reason that the price was so low, was that NASA's budget was based on the economies of scale of their planned 70 launches a year! Of course, that schedule never came true and as the program went on, the estimates were lowered to 60 launches, then 50, then 40, then 20 and in fact were actually only about 6 a year. In 1983 there were only 4 shuttle flights, 1984 there were 5 and in 1985 (when our K-2 las launched) there were 9 -- the most in any one year in the history of the program. RCA benefited from two cheap launches because of NASA's pie in the sky book keeping.

At the time RCA signed up for the shuttle launches, they did not know which satellite would be manifested for their two launches. As the satellite business developed, it became apparent that there was a big market for Medium Power Ku Band spacecraft. Ku Band frequencies are not shared with terrestrial microwave so there are not the interference and earth station siting problems experienced with C-Band services. Because of the higher frequencies and higher spacecraft transmitting power, much smaller dishes could be used on the ground. These advantages opened new markets including TV networking, TV news gathering, radio programming distribution and VSAT Networks. VSAT stands for Very Small Aperture Terminals and refers to the small dishes often seen on the roofs of gas stations and small businesses. VSAT networks can contain many hundreds of digital terminals. Americom's plan was to supply satellite bandwidth for these networks but not to get involved in operating VSAT networks themselves. Unfortunately, from my standpoint, Americom ultimately got

---

putting the KH-12 into a polar orbit that could not be done from Cape Canaveral. The facilities were abandoned before completion and the shuttle never launched a KH-12. The KH-12's were launched on Titan-4 expendable launch vehicles.

pulled into what is a complex and messy business -- but more of that later.

Americom elected to use their two shuttle slots to launch two Astro Series 4000 Ku Band spacecraft. The contract with Astro was for K-1 and K-2 to be launched on the shuttle. A third spacecraft, K-3, would be the ground spare in case of failure of K-1 or K-2. These spacecraft would be larger but not much heavier than the C-Band birds. The communications receiver FET preamplifiers were thermoelectrically cooled to obtain a 2dB noise figure -- the first cooled preamp in a communication satellite. The payload was sixteen 45-Watt, 54Mhz bandwidth, TWTA transponders. Eight transponders were in the horizontal polarization and eight in vertical with an 11 for 8 TWTA redundancy scheme. The larger number of spare TWTA's was because the 45-watt tubes were not considered to be as reliable as lower power tubes. The four panel solar array contained 280 sq ft of cells providing 3600 Watts of electric power at the beginning of life. There were two Nickel-Hydrogen batteries providing a 150 ampere-hour capacity. Because of the enormous amount of heat generated by the payload, the south face communications panel contained ammonia-filled heat pipes to spread the heat out across the OSR radiator. The propulsion system had four (13, 14, 15 & 16) Electrically Heated Thrusters (EHT) on the North face to be used for North/South (N/S) Maneuvers. Although the EHT's did not have as much thrust as the conventional thrusters they were much more fuel efficient and thus lengthened spacecraft life. The solar array shaft was located in the center of the North panel with the thrusters arranged symmetrically around it. N/S Maneuvers could be run with the solar array moving in its normal position with minimal attitude effects from the thruster plumes. Some of the electronic boxes in earlier Satcoms were replaced with an On-Board Computer (OBC). The PCM telemetry system extensively subcommutated the telemetry frames providing much more data particularly for the thermal subsystem.

The two new spacecraft also required additional ground station equipment. Three 9-Meter K-Band antennas and their associated command transmitters and downlink equipment were installed at Vernon Valley and South Mountain. In addition, an HP-1000-A900 Video computer with associated disk drives and a Versatec Plotter were added to the TT&C equipment to provide improved telemetry data display.

As K-1 and K-2 were being built at Astro, the recently approved facilities expansion was under way at Vernon Valley including a new two story TT&C building up the hill from the existing earth station.   As the new building was nearing completion, I was heavily involved with Duane McMillen my VV TT&C Manager in developing plans for the move of the TT&C facility to its new home.   I was concerned that the station downtime be minimized as much as possible. Our TT&C Engineering group had estimated that it would take several days to make the move. Duane and I felt that was totally unacceptable and that with careful planning we could do it in considerably less time. The plan was that South Mountain would fly all of the birds while the Vernon TT&C station was picked up and moved up the hill -- easier said than done. It was true that one station could control all of the spacecraft but there were potential single point failures that could possibly take South Mountain down.   We were trying to put together a transition plan that would involve the minimum down time for Vernon Valley

New fire code required that all of the underfloor cables now be fire resistant plenum cables so all of our existing cables had to be replaced.   Thus we could install all of the replacement cables in the new building long before the equipment move.   All cables were checked for type and connectors against system interconnection diagrams.   The exact placement of each piece of equipment was marked on the floor and cable and air conditioning holes were cut in the floor tiles.   Cables were laid under the floor and carefully tagged and marked. Uninterruptible power was run to each rack position. The existing TT&C racks were unbolted from the floor and one another. Racks for new equipment were installed in the new building. Non-critical ancillary and peripheral equipment was relocated to the new building in the days before the move. Electronic cquipment movers had been hired for the move day. Duane and TP Tubbs our Computer Tech had designated the exact order that the equipment racks were to be moved. The first units to go were the most critical or the ones requiring the most time for installation.

We had planned spacecraft maneuver and ranging schedules so that none of the 7 spacecraft would require a maneuver for at least four days. On the day of the move we had every available RCA technician in the station on overtime plus a couple of contract technicians. We had Hewlett Packard computer techs in both Vernon Valley and South Mountain to assist in case of computer problems. The movers had the truck backed up to the door, their equipment was in place and they were ready to start. The analysts confirmed that all spacecraft appeared to be operating normally. I called Ray Balon, the South Mountain TT&C Manager and confirmed with him that they were ready to take all of the birds. I told Duane, "Transfer your birds to South Mountain". When South Mountain confirmed that they had them, I turned to Duane "Power down--GO". And go we did. Our engineering group in Princeton had estimated it would take 3 days of down time for the move. Vernon Valley TT&C was back up in its new quarters and flying its birds in 16-hours!

Meanwhile, we were learning about the vagaries of launching K-1 and K-2 from the shuttle. The shuttle operating altitude with a communication satellite payload is only about 225 miles above the earth -- far from geosynchronous. This means that the spacecraft must be mated to a rocket motor equivalent to the Delta 3rd stage to put it into Geosynchronous Transfer Orbit (GTO). RCA elected to use the PAM-D2, a slightly larger version of the PAM-D that we had used as Delta third stage for the F-3R and F4 launches. This was the first flight for this $10 million booster. The PAM's deployable (expendable) stage consists of a spin-stabilized, solid-fueled rocket motor; a payload attach fitting to mate with the unmanned spacecraft; and the necessary timing, sequencing, power and control assemblies. The reusable airborne support equipment consisted of the cradle structure for mounting the deployable system in the Space Shuttle orbiter payload bay. There is also a spin system to provide the stabilizing rotation; a separation system to release and deploy the stage and the spacecraft; and the necessary avionics to control, monitor and power the system. As noted before we had spinning wobble problems with the PAM-D on the F3R and F4 launches caused by the exhaust nozzle eroding during burn. We were understandably a bit nervous about PAM performance based on that experience. We became really concerned about some even more disturbing news about PAM launches from the Space Shuttle.

On February 3, 1984, about a year before our K-Band shuttle launches, Hughes was to launch two communication spacecraft from Space Shuttle flight STS-41B. Westar-6 and Palapa B-2, were both Hughes HS-376 Communications Satellites. Westar-6 was launched first and deployed normally from the shuttle bay. Forty-five minutes later, the PAM-D ignited normally but only burned for a short time before sputtering out. After a technical investigation and discussion on the ground it was decided that the problem was unique to the Westar PAM and it was recommended that Palapa B-2 be launched. The Indonesian government concurred and three days after Westar-6, Palapa was deployed from the shuttle bay, and its PAM faltered after only a few seconds of burn. The PAM-D's on both spacecraft did not burn for for anywhere near the required time resulting in neither spacecraft being near the desired Geosynchronous Transfer Orbit. Both spacecraft were stranded in 170 by 720 mile orbits at about 52 degree inclination. You can imagine our concern about what this meant for our PAM-dependent launches. We were also concerned about what these failures would do to launch insurance rates -- they doubled almost immediately.

The spacecraft insurers for Westar-6 and Palapa B-2 decided that they would try to rescue the two derelict satellites to recover some of the costs they paid out to the owners. The two insurers, Merritt Syndicates, Ltd. and International Technology Underwriters, had paid the owners $180 million --$105 million to Western Union and $75 million to Indonesia. Now they paid NASA $5.5 million for Space Shuttle services for the rescue mission costs. They also entered into a $5 million contract with Hughes for rescue equipment and for maneuvering the two spacecraft into rescue orbits where they could be recovered by the shuttle. The insurers financed the rescue mission with the hope of recovering about $50 million by refurbishing and reselling the two recovered spacecraft.

The satellite rescue was engineered by Hughes Chief Technologist Jeremiah Salvatore. Jerry Salvatore was a brilliant, confident, wild man. In February 1982, Jerry was in Vernon for the Westar 4 launch. He and I and an Intelsat guy went skiing at the Vernon Valley Ski Area on the morning after the Westar launch. We had a great time skiing. Riding up on the lift, Jerry was telling me how he wanted to fire the Westar-4 AKM on 3rd apogee instead of the normal 7th apogee. If he did this, he said, he could go back to California two days earlier. He laid out a pretty compelling technical scenario. Evidently the Westar folks weren't quite as courageous as Jerry. I encountered Jerry in the George Inn in Vernon two days later and said that I was surprised to see him. Jerry said the Westar people had no guts and they fired the AKM as normal on the 7th apogee.

Although they were stranded, Westar 6 and Palapa B2 were both operating normally, they were transmitting telemetry and could be commanded. First Jerry's team sent commands to jettison the failed PAM-D module. They then fired the 5 lb South face thrusters on both birds to circularize the orbits at 650 miles. After some ranging and and orbit determinations, a second two-part maneuver brought them both down to 224 miles -- now they were within the shuttle's maneuvering altitude. Space Shuttle flight STS-51-A lifted off from Cape Canaveral on November 8, 1984 at 4:15PM. On the second day after it got into orbit, it successfully deployed two communication satellites, LeaseSat1 and Anik D2. The shuttle crew then started maneuvering the shuttle to chase down the two errant satellites. Ground commands from the Hughes control center had slowed the spacecraft rotation from 100 RPM to 1 RPM. On the fifth day as they approached the Palapa spacecraft, Astronauts Allen and Gardner donned pressure suits and performed an Extra-Vehicular Activity (EVA). They captured the satellite with a device known as a "Stinger," by inserting it into Palapa's apogee rocket nozzle. Using their backpack thrusters and the Canadarm, they then moved the spacecraft into the shuttle bay. They followed the same procedure for Westar 6. The satellites returned to earth when flight STS-51-A landed on runway 15 at Kennedy Space Center. The two HS-376 spacecraft were completely reconditioned at Hughes El Segundo, CA facility. In their nine months in low earth orbit both spacecraft were hit by thousands of micrometeorites that punctured holes in the solar cells. All of the damaged cells had to be replaced. Not knowing what damage the "Stinger" tool may have done to the Apogee Kick Motor (AKM) rockets, the AKM nozzles were replaced. At first Hughes resisted refurbishing the satellites because they would rather sell new birds as replacements. Finally under pressure from the Indonesian government, who were a major Hughes customer, they agreed to the refurbishment.

The Indonesian government had contracted for an "early bird" special on the shuttle for Palapa B-2 similar to the one that Americom had for their K-2 launch. Under the terms of that deal, NASA guaranteed a free shuttle ride if the original launch failed. A California company, Sattel Technologies, had bought B-2 from the insurance company before the retrieval with the expectation that they could use the free launch to put B-2 back into orbit. Unfortunately for them, President Reagan issued an executive order forbidding anymore commercial shuttle launches after the Challenger disaster. Sattel ended up contracting for a $50 million Delta launch to put the refurbished Palapa B-2R into orbit.

Westar 6 was sold to AsiaSat Telecommunications Co. and became AsiaSat-1. AsiaSat-1 was launched from Xichang, China on a Long March 3 launch vehicle on July 4, 1990. I observed AsiaSat-1 operations after I was retired and doing consulting. I was in Hong Kong, working for Martin Marietta at the AsiaSat Control Center, supporting the launch of the Astro-built AsiaSat-2 in 1995. AsiaSat was very pleased with their "previously owned" spacecraft.

Astronaut Dale Gardner moving Westar 6 to the shuttle using his backpack thrusters. *NASA Photo*

Coincidentally Jeremiah Salvatore enters the picture once more with AsiaSat. AsiaSat-3, a Hughes-601 three axis spacecraft was launched on Christmas Day 1998 on a Russian Proton launch vehicle. The Proton fourth stage failed leaving AsiaSat-3 in a useless orbit around the earth. AsiaSat-3's insurers declared it a total loss. Hughes Space and Communications Corp. bought the derelict spacecraft from the insurers. Jerry Salvatore and his team flew the bird twice around the earth and then around the moon and used the bird's thrusters and the moon's gravity to slingshot it into a geosynchronous orbit and parked it over the Pacific Ocean. The recovered Hughes HS-601 bird was ultimately bought by PanAmSat.

One of the negatives of launching from the shuttle was that the entire communications satellite payload had to be "man-rated". That means that nothing can present a hazard to the on-board astronauts. One of the major problems is that communication satellites in launch configuration contain quite a few pyrotechnics (explosives). These are used to cut the shear tie cables holding the folded solar arrays, release "belly bands" around the arrays, separate the

212

PAM from the spacecraft, unlock the MWA pivots and operate fuel system valves. The circuits for each of these explosives have separate ARM and FIRE switches, however, that is not considered sufficiently safe for "man rating". Another box had to be added to the spacecraft containing another level of power isolation for these circuits. This added weight and complexity to the K-1 and K-2 spacecraft and meant that we carried a little less fuel.

An editorial note: *I would never advocate anything that would compromise the safety of astronauts, however, I developed a healthy skepticism of the "man rated" requirement while working for RCA Service Co. at Goddard Space Flight Center (GSFC) from 1965 to 1974. After the Apollo-1 fire, NASA launched a major serious drive to correct any and all astronaut safety issues and the US Congress opened the coffers. Unfortunately with this money available, many in NASA saw an opportunity to fund pet projects by declaring them "man rated". For instance, in GSFC Building 3, in the name of "man rated", NASA built a large, three tiered, 15 console Manned Flight Network Support Team (NST) room surrounded by a nest of offices. Its supposed mission was to help technicians at the world-wide Manned Flight Network tracking stations trouble shoot equipment problems. The lie was exposed by the fact that the manned Apollo launches resumed and in fact achieved the moon landing objective (Apollo 11) before the Manned Flight NST ever became operational. No one ever expressed any safety concerns for those missions over the lack of an NST. The Manned Flight Network NASA folks and their contractors finally got to sit at their fancy consoles for Apollo 15, 16 and 17 that ended the program. That was only one phony man-rated project -- there were others. The reason that example particularly galled me was because I was responsible for the STADAN (Scientific Satellite Network) NST. In my NST, I had two technical guys sitting at desks and a teletype operator and his machine in a small room on the second floor of Building 17 (the bank building). Apollo had 2 or 3 one week missions a year. The STADAN NST supported a 15 station network tracking over 50 spacecraft 24 hours a day 365 days a year. The final irony was that at the end of Apollo, it was decided to "merge the networks' and they invited my people to move into two console positions in the back row of the Manned flight NST -- the rest of the 15 consoles were empty!*

Another negative to launching on the shuttle was the length of the payload launch flow at the Cape. For a Delta launch, the spacecraft is delivered to the Cape and four weeks later it is launched. For a shuttle launch, the spacecraft was delivered to the Cape and twelve weeks later it is launched! That meant that an additional two months had to be added to the project time from procurement of long lead time items to launch. I guess that we were all glad to be getting a cheap launch and there is also something romantic about manned spaceflight but it also seemed like a lot of trouble.

For some reason or other K-2 was manifested on STS-61B in November 1985 and K-1 was manifested after K-2 on STS-61-C in January 1986. As noted earlier, NASA had guaranteed RCA a replacement shuttle mission if the K2 mission failed. For that reason, and the extremely high insurance rates, RCA did not insure the launch -- they put some money in reserve to replace the spacecraft. Those rules did not apply to the K1 launch, however, which RCA insured. Because of the previous problems with shuttle/PAM launches, the insurance cost RCA 32% of the insured value. By contrast, later launches from French Guiana on the Ariane launch vehicle cost as little as 17%. As the K-2 launch neared, Astro made arrangements for the NASA Astronauts to visit the Astro plant in East Windsor, NJ to learn more about the spacecraft that they were to launch. It was also an opportunity for the RCA folks to meet the astronauts. On the night before the plant visit, my boss Joe Schwarze, Director, Space Systems, myself and our Spacecraft Engineering Manager Joe Elko had dinner with the astronauts at Goodtime Charlie's in Kingston, NJ near Astro. It was in a private room with wine and cheese before dinner and wine with dinner so we were a pretty convivial group. It was a very enjoyable evening with a lot of very interesting conversation. Brian O'Connor, a marine colonel was on my right and we talked about the Corps and my sons imminent induction into it. Mary Cleave on my left was a very interesting person. With a PhD in Civil Engineering she was also an avid scuba diver, private pilot and ski instructor. Mary would be the one that actually launched our K-2 bird out of the shuttle. Over dessert, Mary announced that what we needed was a round of Yukon Jack -- we all agreed and toasted to our success.

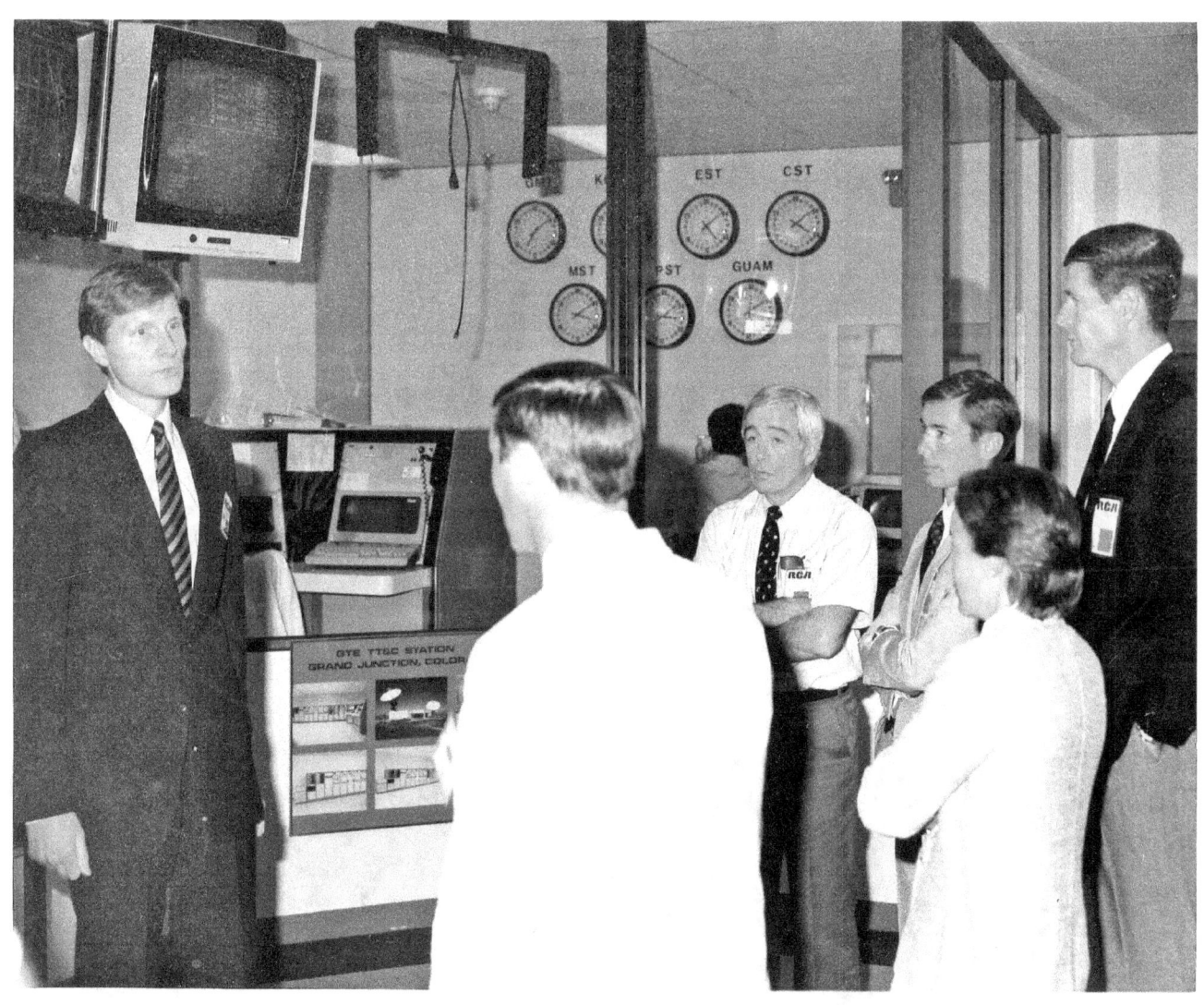

K-2 Mission Astronauts visiting RCA Astro Spacecraft Operations Center (ASOC)   L to
R: RCA Mission Director Dick Hroussovsky, Astronaut Jerry Ross, the author, Astronauts
Brewster Shaw, Mary Cleave and Brian O'Connor - *RCA Astro-Electronics Division Photo*

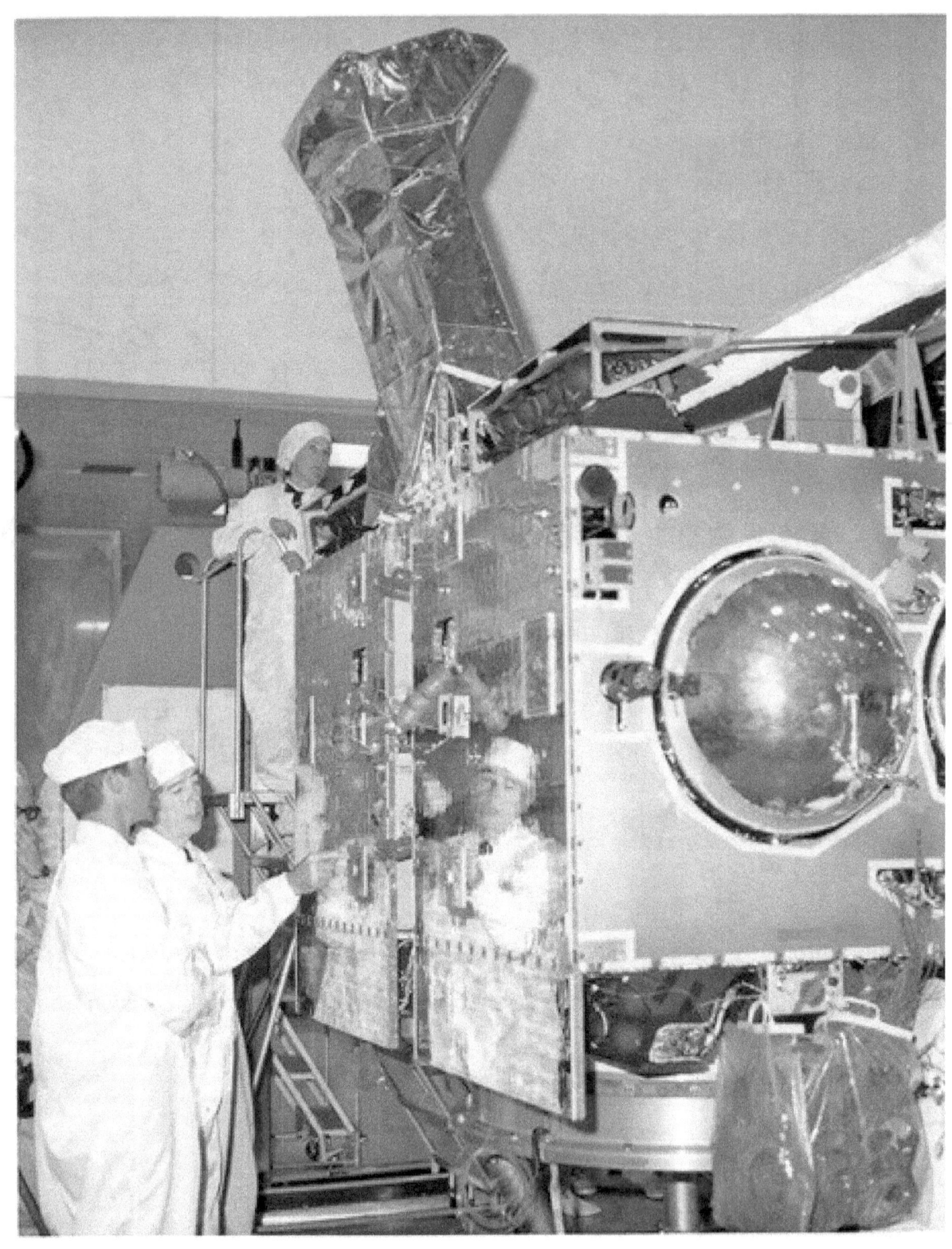

K-2 Mission Astronauts examining the spacecraft in Astro White Room. The author is pointing out one of the Electrically Heated Thrusters (EHT) to Brewster Shaw. Mary Cleave is on the ladder examining the gold mylar covered RF Feed Assembly. - *RCA Astro-Electronics Division Photo*

The Space Shuttle Atlantis lifted off from Pad A, Launch Complex 39, KSC at 7:29PM EST on November 26, 1985 carrying RCA's Satcom K-2 to orbit. It was both the second flight for Atlantis and the second night launch in the shuttle program. In preparation for the launch, the sun shield was retracted from K-2. The sunshield protects the spacecraft from getting excessively hot when the shuttle bay doors are open. Mary Cleave then activated the PAM spin table that spins the spacecraft up to 60 RPM. On November 28 (Thanksgiving) at 4:29 PM EST when about 210 miles above the earth, Mary launched Satcom K-2 from the shuttle bay. Springs pushed the spinning spacecraft out of the bay. The PAM 45-minute timer started and the shuttle started to move away from the spinning spacecraft and its PAM-D2 booster. The timer allows the shuttle to get well clear before the PAM Thiokol rocket motor fires. The motor fired and then K-2 separated from the PAM on its way to Geosynchronous Transfer Orbit (GTO). Their work done, the crew settled down for thanksgiving dinner. The seven crew members had a dinner of chicken consomme, smoked turkey, cranberry sauce, green beans, corn, pasta, butter cookies and lemonade. About 8:30 P.M., as the shuttle sailed 230 miles over Mexico, television views were beamed down showing the crew preparing their Thanksgiving meal. Floating in the shuttle's cramped middeck, the space chefs had to stick their food trays to Velcro strips on the wall to keep them from floating off. In the following days the crew completed their other experiments including the construction of a tower in the shuttle bay. The shuttle landing was at Edwards AFB, at 16:33 EST on 3 December 1985, after a mission duration of 6 days, 21 hrs, and 5 minutes.

Satcom K-2 on its PAM-D2 booster emerging from the shuttle bay. The dark panel on the right side of the spacecraft is the folded up solar array. The sun shield for another communication satellite in the bay can be seen in the foreground. - *NASA Photo*

Over the next three days K-2 was controlled by the ASOC and tracked by the Astro stations at Alpha, NJ and Guam, Marianas. After the AKM was successfully fired, the Americom stations at Vernon Valley and South Mountain tracked K-2, processed telemetry and took ranging files as it drifted toward its station at 81 degrees West Longitude.

For the K-1 shuttle mission, an RCA Astro engineer, Bob Cenker, would be in the crew as a Payload Specialist. Bill Nelson, Congressman (now Senator) from Florida was the second Payload Specialist in the crew. I had known Bob Cenker from the early days of the Satcom program. At that time he was a young guy, not far removed from Penn State where he had gotten a Bachelors and Masters in Aerospace Engineering. He subsequently had gotten a Masters in Electronic Engineering from Rutgers. He was part of the Astro Attitude Control group for our F-1 and F-2 Spacecraft launches and we got to know one another quite well. I found Bob to be a bright guy and he was a big help to me in understanding the complexities of our new spacecraft. Bob was the Integration & Test Manager for our F3R and F4 spacecraft. He was the guy in Astro responsible for getting our birds put together and checked out. I was very happy to see Bob get this great opportunity to go into space.

Bob Cenker, RCA Astro Payload Specialist (Black suit) with crew members of STS-61-B that launched RCA K-2 during a tour of the Astro-Electronics plant. - *RCA Astro-Electronics Division Photo*

The Space Shuttle Columbia Mission STS-61-C to launch K-1 seemed star crossed. This was the first mission for the shuttle Columbia after a two year refurbishment and modernization overhaul. Launch set for Dec. 18, 1985 was delayed one day when additional time was needed to close out the orbiter aft compartment. A launch attempt Dec. 19 was scrubbed at T-14 seconds due to an indication that the right solid rocket booster hydraulic power unit exceeding RPM redline speed limits. (Later determined as false reading.)   Because of the time to repair and the KSC Christmas shutdown, the next launch attempt wasn't made until January 6, 1986.   Launch attempt on Jan. 6 was halted at T-31 seconds due to accidental draining of approximately 4,000 pounds of liquid oxygen from external tank.   The Launch attempt on Jan. 7 was scrubbed at T-9 minutes due to bad weather at both transoceanic abort landing sites (Moron, Spain and Dakar, Senegal). After a two-day delay, launch was set for Jan. 9 and then further delayed due to launch pad liquid oxygen sensor breaking off and lodging in number two main engine prevalve. If not detected, that problem could have caused a catastrophic failure. Launch was reset for Jan. 10 then delayed two days due to heavy rains.   Launch countdown on Jan. 12 proceeded with no delays. The launch finally took place at 6:55 a.m. EST, on January 12 without further problems.   We all breathed a big sigh of relief that after seven aborts, STS-61-C finally got off.

STS-61-C Orbiter Columbia lifting off from KSC. - *NASA Photo*

The primary mission of STS-61-C was launching our K-1 spacecraft.   Unlike some of the earlier shuttle missions our communication spacecraft was the only one on board.   Satcom K-1 was launched from the shuttle bay without incident, the PAM-D2 burned successfully, and the Astro ASOC took control.

Satcom K1 leaving the shuttle bay of STS-61-C. The equipment in the bay is the "Get Away Special Bridge" carrying the scientific experiments to be conducted by the crew. - *NASA Photo*

Not only was the STS-61-C mission difficult to get off the ground, it proved to be difficult getting it back to Earth. It was originally scheduled to land on January 17, but this was brought forward one day because the delays of STS-61-C were causing the next flight, STS-51-L, to be delayed. The landing attempt on January 16 was cancelled because of unfavorable weather at Edwards AFB. Continued bad weather forced another wave-off the following day, January 17. The flight was extended one more day to provide for a landing opportunity at KSC on January 18 — this in order to avoid time lost in an Edwards AFB landing and turnaround. However, bad weather at the KSC landing site resulted in still another wave-off. *Columbia* finally landed at Edwards AFB at 5:59 a.m. PST, on January 18. Mission elapsed time was 6 days, 2 hours, 3 minutes, 51 seconds.

Over the three days after launch from the shuttle, K-1 was controlled by the ASOC and tracked by the Astro stations at Alpha, NJ and Guam, Marianas. After the AKM was successfully fired, the Americom stations at Vernon Valley and South Mountain tracked K-1, processed telemetry and took ranging files as it drifted toward its station at 85 degrees West Longitude.

RCA Americom spacecraft K-1 was the last commercial communication satellite launched by the Space Shuttle. The next Space Shuttle launched after the K-1 mission was the Challenger that ended in disaster. In the aftermath of that disaster, President Reagan announced in August 1986 that the shuttle would no longer carry commercial satellite payloads.

Chapter 17 - "GE - We Bring Good Things to Life"

On December 11, 1985, in the middle of our K-Band launches, GE bought the RCA Corporation for $6.3 Billion. We all thought that this was good for us generally but wondered how it would all work out for Americom. It didn't take long to find out and it turned out to be the best thing that could have happened to Americom. Jack Welch, GE CEO, appointed Kevin Sharer, a rising star in GE to be President of GE Americom. Sharer was a Naval Academy graduate, nuclear submarine officer and had worked at MacKinsey and Co. as a consultant. He joined GE Consulting and came to Americom from there.

Sharer immediately implemented Welch's Work-Out (short for "get the work out") program. At its core, Work-Out is a very simple concept based on the premise that those closest to the work know it best. When the ideas of those people, irrespective of their functions and job titles, are solicited and turned immediately into action, an unstoppable wave of creativity, energy and productivity is unleashed throughout the organization. Welch's dictum was "look at everything that you do -- if it doesn't add value to the product that you are delivering to the customer -- don't do it." The first Americom Work-Out sessions were large 80 -100 people that were then broken up into teams of 6 to 8. Each team decided on something that they thought was a problem and would like to work on. At the end of the day, each team made a presentation to Sharer and the Sr. VP's with recommendations for changes that they would like to see made. Most of the time the team would be told to go ahead and implement the changes. In some cases they were told "this seems like a good idea but we need more info on these specific things." Very few things were turned down. I thought Work-Out was great. We got rid of a lot of useless reports and Mickey Mouse procedures and it definitely got the work out. It was also made very plain to all of us that we were to solve our problems by talking to one another. Cover Your Ass memos would not be tolerated. Later after most of the easy stuff was taken care of, the Work-Out groups got smaller and a lot more intense as they dealt with long standing and difficult organizational and personality problems. It was not easy but it forced all parties to a problem to confront the issues and to talk through a resolution. By and large it worked and the result was a much more open culture.

The Americom organization was flattened out. Welch believed that there should be no more than three layers of management within any organization. We mostly achieved that. I now reported directly to a Vice President and I had more people reporting to me. My responsibilities were expanded to encompass all of the functions at Vernon Valley including the Communications Earth Station and the Network Operations Center.[9] I have to admit that I resisted the reorganization. Walter Braun, Senior VP Engineering and Operations proposed the change to me and I kept going back to him with alternate plans. I felt that I had enough on my plate with Spacecraft Operations and that the Vernon Valley Earth Station had serious technical, organizational and union problems that I could do without. Finally at a meeting in Walter's office, he said "Here's the deal -- take the job or I'll fire you". Yes Walter!

The Vernon Valley Earth Station had suffered all of the technical problems associated with booming business and fast growth. A lot of equipment had been installed, not always in the best way but in the fastest way to get new customers on the air. The downlinks had unnecessary RF power dividers and cables and the uplinks had excessive waveguide runs and complicated switching. The space under the raised floors was filled with abandoned cables and waveguide. If a cable was no longer needed, the quickest solution was to just drop the ends under the floor. In haste, primary and backup equipment had been connected through the same power panel compromising reliability. I started a major clean up campaign to get the equipment reconfigured in the most reliable configuration and to get all the "dead" cables removed from the under the operations room floor. It didn't make me popular with the technicians because I scheduled this activity on the overnight shifts where the techs were usually catching up on their reading or TV programs. We did, however, improve reliability and salvage many hundreds of feet of coaxial cable and waveguide to be used in future

---

9   Walter Braun, Executive VP and a normally a rather humorless individual described my promotion as "King of Vernon Valley".

projects.   We also drastically simplified the downlink RF cabling, removed power dividers and raised signal levels to the receivers.   I also redesigned the uplink RF combiner plates in the HPA Room to increase flexibility, simplify backup switching and insure that we could uplink to all spacecraft in both polarizations if required.   The reconfiguration of the plates was tricky because it required planning and scheduling around active uplink equipment. Because of that and the need to search our system for the filters we needed, the uplink reconfiguration took a lot longer than the downlinks.

The Network Operations Center (NOC) had originally been called the Network Monitoring Center and did just that -- techs watched a wall full of TV's and reported problems they saw to others.   I wanted to turn it into a real operations center responsible for controlling access to the satellites and to assist in resolution of problems.   I had the NOC test, switching and display equipment reconfigured so that the operators had the flexibility to quickly look at and investigate problems.   Twelve small monitors could now be switched to look at the traffic in all twelve transponders in either polarization on any the spacecraft.   We reconfigured the switching system so that when a large TV monitor was switched to a channel to investigate a problem, a TV Monitor in the console, a Waveform Monitor and a Video Test Set were also connected to that channel so that quantitative measurements could be made and reported to the customer. This also meant an intensive period of generating operations procedures and training personnel.   Slowly but surely the operations were becoming a more customer focussed quality oriented environment.   It wasn't easy - I tangled with the union a lot, however, they came around as they slowly realized that I was trying to improve the operation and treat everyone fairly.

It didn't all go smoothly. I had two groups of technicians in Vernon Valley; Communication Technicians and NOC Technicians.   I wanted to combine the two groups so that all the techs could either work in the earth station or in the Network Operations Center (NOC).   The NOC techs weren't too happy about the idea but were not resisting it. The Communication techs, however, were in an "over my dead body" mode. It took a number of meetings with the union shop steward and business agent, a couple of Work-Out sessions with the techs and the establishment of a training program but we were finally able to combine the groups.. I had to fire two union technicians; one for repeatedly sleeping on the night shift despite warning letters and one for deliberately falsifying a time card. Despite comprehensive documentation of the details of both incidents, the union fought tooth and nail, however, I was able to prevail. The techs now realized that management would help them to do their job but would not tolerate malfeasance or malingering. It didn't make me happy though -- I always looked on terminations as a failure.

When the Vernon Valley Earth Station Manager resigned I brought in the manager of the Americom Lake Geneva, Wisconsin Earth Station that served Chicago. He came recommended and had a reputation for doing a good job at Lake Geneva, however, within a couple of months at Vernon I realized that I had made a mistake. He came to work in a three piece suit, walked around the station all day, provided little direction to the techs and didn't accomplish much. I had a couple of sit downs with the guy followed up with a letters documenting our discussion and agreements. There was no improvement and in fact things got worse. Finally, I to put the NOC Manager in charge of both the earth station and the NOC and demoted the earth station manager to Station Engineer. Unfortunately, he didn't perform well in that job either and I ultimately ended up terminating him. The only good that came out of all of this was that I now had both the earth station and NOC management and the technicians in a consolidated organization. It was still not a happy marriage and took continued training and mentoring of the technicians but it was working.

GE emphasized customer service and we were encouraged to make sure that all people dealing with customers trained those people in the fundamentals of a customer oriented business. The Vernon Valley technicians regularly talked to customer technicians and engineers. No one could transmit a signal to an Americom spacecraft unless cleared through Vernon Valley. Many of the customer technicians were not well trained and my techs were quick to note that and treated them derisively and sarcastically while talking to them on the phone. It was hard to convince the Americom techs that they should treat the customer personnel politely and try to help them. This was particularly true when the customer had difficulty alining their antenna to come up on the satellite. When I suggested that the techs should try to work with the customers, the usual response from the techs was "they're a bunch of idiots -- they should be fired". I tried to point out that it was the customers that indirectly paid their salaries -- didn't seem to help. One refreshing thing about GE was that the Corporate Staff was a big help with some of our problems. GE staff people arranged for Customer Service training for the techs. That and continued mentoring by myself and Dave Gardner the Earth Station/NOC Manager gradually improved things.

The original Network Monitoring Center at Vernon Valley. In the foreground from left to right, George Kowal, Ronald Korman, Steven Nieschower, Richard Hanf and Richard Hanna. Behind the console are Alice Scanlon, Bryan Louie and Oleg Matiash - *RCA Americom Photo.*

Sharer put a big emphasis on Customer Service and directed that all of the management (even those of us in Operations) had to be in contact with customers. I was made Technical Liasion to our NBC customer which was very interesting. I attended the monthly Operations Meeting at 30 Rock or Burbank and worked closely with the NBC Chief Engineer, Charlie Jablonski. He was a very smart guy and had won four "Emmys" for his technical achievements during NBC's coverage of the Olympics. I also attended the annual Affiliate Chief Engineer Meetings that were held in either Chicago or St. Louis. The Chief Engineers from the ten NBC wholly-owned stations and about 200 affiliate stations gathered for what was primarily a technical conference and to discuss network problems. I certainly got a better understanding of how NBC operated and a much better understanding of their problems.

I was also the Americom technical liasion to IDB Communications, a company originally formed by Jeffrey Sudikoff to do radio broadcasts of rock concerts. Sudikoff was noted in the industry for agreeing to do almost anything asked without worrying too much about the technical details. IDB owned the Los Angeles Teleport on Washington Boulevard in Culver City and they had recently acquired rights to operate the Staten Island Teleport serving New York City. I ended up dealing with problems from both of their locations.   One of our customers, a California regional sports network operator, LA Sports, had contracted with IDB to put their programming up on F-4. IDB proposed a lower rate for uplink services if LA Sports would allow IDB to add four FM radio subcarriers to their F-4 transponder. This was proven technology and should work fine and not degrade the the video. Pretty soon the NOC started getting calls from LA Sports stating that their video had a moire pattern in the picture -- IDB said that it was caused by the satellite.   The NOC techs checked the downlink on a spectrum analyzer and they could clearly see that the four IDB audio subcarriers were not set up correctly and were interfering with the video carrier. We told our customer what the problem was, however, IDB continued to deny that it was their problem. Before I left for LA, I had our engineering group do a transponder loading analysis and write up the correct power level numbers for each of the subcarriers. I met with the IDB Chief Engineer at Culver City. We discussed the problem and I told him that it didn't look to us like the subcarriers were set up correctly. He denied this and I showed him the paper with the Americom engineer's calculations. He became furious at that point, balled up the paper and threw it in the trash can and yelled at me "I know how to set up subcarriers and I don't need any help from you guys". The next day, back in New Jersey, I called LA Sports and told them what had happened at IDB. The LA Sports engineer was a happy camper -- he said that today his picture looked perfect, no moire. I went into the NOC and looked at the IDB downlink -- sure enough, the subcarriers were setup perfectly. I guess the IDB Chief Engineer retrieved the paper from the wastebasket.

Sometime later a more serious problem occurred. Every afternoon between four and four-thirty PM the NOC was reporting intermittent interference to our C-Band spacecraft programming. Following our standard procedure, the NOC techs called all uplinks and teleports, including IDB Staten Island, asking if they had a carrier up on our birds. All, including IDB, denied it was them. We had also had gotten calls from Westar and Hughes -- they were having the same interference problem that we were having around 4PM. We and Westar and Hughes came up with a plan to divide up the calling so that we could cover all the uplinks while the interference was still going on. Guess what? We finally determined that it was coming out of the IDB Staten Island Teleport. A technician who I had fired was now working at IDB. One of the problems we had with this guy was that he spent more time watching TV than he did working. He was on the swing shift at IDB and every afternoon when he came to work he would use one of their antennas to scan all the spacecraft looking for a good program to watch. He never checked to see if there was a carrier being transmitted (which there was) before he moved the antenna. Americom, Westar and Hughes demanded that IDB fire the guy, which they did.

Chapter 18 - K-Band Business

The K Band Spacecraft used different and much higher frequencies to carry communications between ground facilities and the spacecraft. All of the previous Americom spacecraft had used C-Band frequencies -- 6 GHz uplink to the spacecraft and 4 GHz downlink. The principal disadvantage with C-Band was that these same frequencies were also used by terrestrial microwave systems. Because of this, satellite earth station antennas had to be sited carefully to avoid interference from microwave systems. Finding electronically quiet locations in urban areas was particularly difficult -- in some cases grounded wire mesh shields had to be placed around the sides of the satellite dish to prevent interference.

The Ku Band spacecraft used a 14GHz uplink to the and 12GHz downlink. These frequencies are not shared with any other communication service so there are no interference problems. Also much smaller antennas can be used at these frequencies making it easier to locate them on roofs. A disadvantage of K-Band is that these signals are attenuated more by rainfall and reception is affected more by snow or ice in the dish antenna.

Once the K-birds were on-station, we conducted our usual on-orbit tests. Both spacecraft checked out well. The communication performance was outstanding. The received downlink levels were higher than expected and television signals exhibited better quality (luminance weighted signal to noise) than video transmitted in terrestrial microwave radio links. It didn't take long for both birds to fill up. It was not long, however, before we also had a full blown operations crisis with two of our major customers on K-2, NBC and AT&T.

NBC leased all eight transponders on the "Even" side of K-2. Under contract to NBC, Comsat General Corp. had built a ground network of Ku Band facilities at each NBC affiliate TV station, major uplink facilities at New York and Burbank, a master control center called

Skypath at 30 Rockefeller Plaza in Manhattan and a secondary Skypath at NBC Burbank, CA. NBC's New York uplink plans were complicated by the fact that "30 Rock" is on the National Register of Historic Places. Because of those restrictions, they couldn't put any antennas on the roof no less two 9-Meter Ku Band dishes and several equipment shelters. NBC made arrangements to put their uplink facility on the roof of the Celanese Building about a block away. Of course that meant that they would have to run cables under the streets of Manhattan to interconnect the two facilities -- which made everyone in NBC rather nervous. Wanting to hedge their bets NBC elected to run one set of cables under the street in the Manhattan Cable right-of-way and another set of cables through subway tunnels. Lucky they did. In 1987, there was some major construction work on the Avenue of the Americas near 50th Street. The pavement had been spray paint marked to show where all of the underground facilities were located. All concerned were standing on the corner checking the markings. The guys in hard hats were measuring the pavement and looking at the drawings. Murphys Law being in full effect, however, as soon as the digging started, they cut NBC's cables. Oh well -- this is why you have redundancy.

NBC's ground system allowed them to make optimum use of their eight transponders. Each of the affiliate TV stations had a primary Ku Band system with a 3.5 Meter dish and a secondary system with a 2.5 Meter dish. All of the uplinks and downlinks at the affiliate TV stations and those at Burbank and New York could be controlled from Skypath either manually or under computer control. One K-2 transponder carried the Network East feed for the Eastern and Central time zones and another carried the Network West feed for the Pacific time zone. The Mountain time zone stations generally took a combination of East and West feeds. Two transponders were dedicated to Satellite News Gathering (SNG) feeds. Affiliate stations were scheduled in five minute increments to playback taped news. The feeds were run remotely from Skypath with their computer bringing up the carrier, rolling the tape and bringing down the carrier automatically. Different stations were uplinked automatically every five minutes "untouched by human hands" -- pretty neat. A couple of other transponders were used for SNG from satellite trucks. All of the trucks had to contact the Americom Network Operations Center (NOC) before they could transmit to the satellite. An Americom technician monitored the truck's uplink carrier to assure that their cross polarization isolation was within specs and that they did not exceed allowable power.

The NBC Ku-Band Earth Station on the roof of the Celanese Building. Directly behind the two 9-Meter dish antennas are the white modular shelters containing transmitters and receiving equipment. The modular shelter between the antennas contained video/audio processing and monitoring equipment and interface and conditioning equipment for the cable link to 30 Rock and a maintenance shop. The smaller modular shelter behind the elevator equipment house contained emergency power generators and an Uninterruptible Power Systems (UPS) equipment. The intricate and massive steel I-beam structure needed to support all the equipment on the roof can be seen around the dishes - *Google Earth Photo*

AT&T had two channels in the vertical polarization for their Skynet Star Network VSAT business. AT&T's two channels were opposite from two of NBC's channels in the horizontal polarization. That was not an optimum channel assignment. AT&T had literally hundreds of tiny VSAT carriers in their transponders. These were primarily data carriers from small

233

antennas on the roofs of about 500 gas stations, stores and office buildings. On the opposite polarization, NBC had saturated TV signals using all the power in each channel.   The isolation between the two polarizations was about 34dB so the NBC signals should not interfere with the AT&T VSAT traffic.   If, however, anything such as heavy rain or antenna misalignment changed the polarization and reduced this isolation, the errant NBC signals leaking into AT&T's transponder could blow the tiny VSAT carriers off the air.   And guess what -- they did.   One morning, I got a call from an irate AT&T manager charging that during the night, one of the NBC stations had come up with bad cross-polarization isolation and had taken down the AT&T traffic for about five minutes.   I immediately contacted NBC Skypath in New York and gave them the time of the incident.   Skypath said that TV Station KUTV Salt Lake City was up on K2 at that time. They promised to have Comsat (who built and maintained their network) send someone out immediately to check out the KUTV uplink.   Later that afternoon, the Comsat tech at KUTV called the Americom Network Operations Center (NOC) at Vernon Valley and the Comsat and Americom techs verified that the KUTV antenna operation was within specs.   A puzzlement -- AT&T definitely had interference but the suspect uplink tested out OK.

Over the next few weeks, the incidents of NBC uplink stations interfering with AT&T VSAT's continued. Fortunately, they usually only lasted about five minutes.   Short or not, however, I was getting it from both sides.   AT&T was furious because their customers were losing data and were threatening to leave AT&T.   NBC was furious because Americom was making them drop interfering carriers and they were losing TV traffic.   The Comsat/NBC Skypath system was a completely automated system. NBC affiliate stations could buy five minute blocks of satellite time to uplink news and sports feeds to New York or Burbank or to other affiliate stations. They scheduled the feed with Skypath in New York and at the appointed time Skypath automatically turned on their uplink transmitter, started their tape player to run the feed and then shut everything down as the feed was complete.   In the late afternoon during the prime news gathering time, you could watch the K2 NBC affiliate transponder and see the

234

continuous parade of five minute feeds from different stations streaming into the network. We had our techs monitoring those feeds watching to see if any of the affiliates had poor cross-pol. Never happened during the day -- the interference with AT&T always seemed to occur at night.

It was happening enough. however, that we had a serious problem with our customers AT&T and NBC and it was always the same story. AT&T would complain, we would have NBC drop the offending carrier and Comsat would send a tech out to check the antenna. Our techs would monitor the Comsat tests and the antenna would check out OK. The whole thing became a huge finger pointing exercise. I was the Americom Technical Liasion to NBC and participated in all the technical meetings at NBC. It started out with just the Americom, NBC, Comsat and AT&T technical people involved in the meetings, however, it soon escalated as each side brought in"experts". The Comsat and AT&T position was that the K2 satellite was at fault and of course Americom's position was that the Comsat earth stations were the culprit. Comsat's retort was "You checked our earth stations and they were OK". NBC didn't take any sides -- they just wanted the damn problem fixed so they didn't have to drop feeds. As the meetings went on, Americom's Engineering VP, NBC's Operations VP and a VP level guy from Comsat Labs got into long technical arguments in the nitty-gritty of satellite radio link design -- and we still weren't making any progress. For the next meeting, I volunteered to present data on the history of all of the interference incidents.

As I put together my Power Point presentation it quickly became apparent that a small number of the NBC affiliate stations caused most of the problems. I made a slide with a bar graph showing the number of incidents by station. At the next meeting at NBC, when I put my bar graph slide up, I was pointing out the repeat offenders that were clustered to the left side of the chart. One of the NBC technicians at the back of the room piped up and said "Those are the Phase 1 stations". Someone else said "yeah and they have the original design antenna feeds". The meeting erupted into a jumble of conversations. As soon as order was restored, I pointed out that KUTV in Salt Lake City was the worst offender and suggested that Comsat pull the

KUTV antenna feed and send it back to the manufacturer Vertex RSI.   The next day Comsat had KUTV's feed on its way to Vertex.

The Vertex 3.5 Meter Delta Gain antenna used by Comsat in the NBC network was a unique very efficient design.   It was designed by a good friend of mine Helmut E. Schwarz.   Helmut, who was German, started out working for the famed antenna designer Peter Foldes at RCA Montreal.   Peter Foldes's Montreal team designed the overlapping petal, polarization frequency reuse antenna for the Satcom spacecraft 1 through 4.   The Montreal team also designed the monpulse tracking feed for Americom's 13 Meter Datron Tracking Antennas. Shortly before the F1 launch, Helmut left RCA Montreal and joined Americom.

I had met Helmut during the installation and checkout of the Vernon Valley Datron 13-Meter Tracking Antenna.   Shortly after Helmut joined Americom he came into my office at Vernon Valley and said "can I have the 13 Meter E-Systems Antenna for a couple of hours?" That new antenna was a real sore spot with me.   The E-Systems people, who installed the antenna, had worked on it for weeks and couldn't get the side lobe performance to meet spec. At this point Americom wouldn't accept the antenna, the E-Systems folks had disappeared and I didn't know when the antenna would be made right.   I said to Helmut "Do you want to test it to see what the problem is?" He said "no I'm going to fix it".   Within a few hours he had adjusted the sub-reflector and the antenna was performing within specs.   I was very impressed to say the least. Helmut was an enormous asset to the Americom engineering team and did outstanding work. He was also an avid skier and he and I skied Hidden Valley occasionally at night after work.

Ultimately Helmut left RCA to join Radiation Systems Inc. (RSI) in Melbourne, FL.   I would see him occasionally when he did work for us and he told me about a unique antenna feed that he had designed and was getting patented,   Unlike the conventional Cassegrain or Gregorian feeds, it did not require a large sub-reflector mounted on a tripod structure above the dish.

His design just included a small sub-reflector built into the end of the feed structure.   It allowed the same performance with a smaller dish because the dish wasn't "shadowed" by the sub-reflector and its supporting structure.   Helmut subsequently left RSI and formed his own company Vertex to manufacture and market his new antenna. The Vertex antenna was selected by Comsat for the NBC Skypath system based on its excellent performance.

A Vertex 3.5 Meter Ku Band antenna as the feed is being installed. The sub-reflector and its support posts can be seen at the left end of the feed. This is either a Phase 1 or 2 feed before the posts were replaced with a round fiberglass support. - *New Era Systems Photo*

When the KUTV feed arrived at Vertex in Texas it was tested and performed well within spec, however, subsequent testing showed that when the feed got cold the cross-pol isolation deteriorated badly. It seemed that in the Phase 1 antennas the supports for the small sub-reflector were made of aluminum. When the feed got cold (at night in Salt Lake City for instance), the aluminum supports contracted and the sub-reflector moved out of focus and cross-pol isolation degraded.   The next day when the antenna was tested, it was warm and the sub-reflector was back in focus and cross-pol was back in spec. Mystery solved. The NBC Phase 2 feeds had stainless steel supports that did not move as much with temperature changes. Later feeds had a fiberglass support that worked even better. All of the NBC Phase 1 feeds were replaced and AT&T's problem disappeared -- and was I glad.

Unfortunately, the problems and finger pointing over the past few months had resulted in a fair amount of hostility and bad feelings in both the Americom and NBC organizations. Even at the working level, our technicians in the Americom Network Operations Center (NOC) and NBC Skypath had been in fierce arguments over who was causing the problem and whether NBC had to take down carriers. As part of the get well program we initiated a technician interchange with NBC. Two technicians from Skypath came to Vernon Valley and worked shift at our consoles and two of our technicians went to New York and were on shift in Skypath. Every week we exchanged two more techs until all the technicians had been through the program. It not only ended the hostility but it actually improved both of our control center operations because of the improved understanding of how both groups worked.

Americom realized that commercial broadcast television stations that already had C-Band receiving systems would be reluctant to invest in a K-Band system. Americom started a Syndication System service that offered a Ku-Band antenna/receiver package for Satcom K-2 to these broadcast stations. By the end of 1985, RCA Americom had commitments from nearly 700 stations to accept the packages. As a result there were a number of other television broadcast customers on K-2, besides NBC.

One of these was the fledgling United States Football League (USFL). The USFL was the first major sports league to utilize satellite technology for national, real time distribution of its schedule. "What this means," USFL Commissioner Harry Usher said, "is that we can arrange for satellite uplinking at virtually any location in the country, even in the most microwave-congested metropolitan areas. Using mobile Ku-Band uplinks at the game site, we'll give our audience real time coverage of our 1986 schedule without the high cost of terrestrial connections to a remote satellite uplink facility."

Cycle Sat was an interesting company that used the satellite to distribute commercials to TV stations. The company was a subsidiary of Winnebago Industries and was located next to their RV factory in Forest City, IA. They transmitted the commercials through K-2 to dedicated satellite receivers and tape decks at client TV stations where they were recorded automatically during the overnight. For TV stations that did not have satellite facilities, Cycle-Sat transmitted the commercials through K-2 to the Cycle Sat tape production facility across the street from FedEx's hub facility in Memphis. They could then take advantage of low shipping rates to TV stations for tapes delivered directly to the FedEx hub. One of the Cycle Sat principals once told me that the bulk of their business was movie commercials. When I questioned why that was, he said "there is not one commercial for a movie -- there is "Coming This Spring", "Coming Soon", "Starting April 15", "Starting Friday" and "Now Showing"". Commercials were also in 15, 30 and 60 second versions. Ultimately Cycle Sat became a $40-million business with 550 network affiliates. They routinely send over 200 commercials a night to their affiliates. If an agency delivered the commercial to Cycle Sat by midnight, they guarantee that it would be delivered to the affiliates by 6AM. That's why you see an ad for an Oscar winning film on the Today Show the morning after the Academy Awards. Quite a business. Evidently their video transmission system was pretty robust too. In 1987, we had an operator error incident during a North/South maneuver on a Sunday afternoon[10] that resulted in K-2 losing lock on the earth. When we got K-2 locked back on the earth it was still nutating pretty badly and our opinion was that it was not suitable for carrying traffic. One of Cycle-Sat's engineers called me and asked if I cared if they started transmitting commercials. He said that they had a large backlog of commercials to send overnight and would like to get started. They had their carrier up in K-2 with color bars and they had checked with some of their affiliates and they were receiving it OK. It took us more than an hour to get K-2 completely settled down and through it all I could see Cycle-Sat on the spectrum analyzer transmitting commercials. Pretty good system.

---

10 - NBC was carrying an NFL Seattle Seahawks game on K-2 when we lost pitch lock. Needless to say NBC was very upset and that upset came down the Americom chain of command very quickly to me.

Another Americom K-2 customer, CONUS in St. Paul MI, was a pioneer in the development of the Ku Band Satellite News Gathering. CONUS Director of Engineering, Ray Conover, designed the first SNG truck in 1984 and named it CONUS1. The system had a K-Band 2.4 Meter dish mounted on the back of a Ford truck. The truck also contained the uplink and downlink electronics and video and audio broadcast equipment. CONUS displayed the truck at TV stations and broadcast conventions and before long almost every TV organization had SNG trucks.

The SNG trucks allowed local TV stations nationwide to provide live coverage. As the use of the SNG trucks grew, CONUS News Service expedited live coverage of events and brokered video footage exchange between stations and networks. They also coordinated coverage of major news events such as hurricanes, earthquakes, political conventions and breaking news. They had regional news bureaus across the US including a full production facility in Washington D.C. Under their contracts, if a TV station gave video footage to CONUS for transmission to another station or network, CONUS then had exchange rights and the footage was available for further distribution. That fact cost NBC an "exclusive" in the crash of United Flight 232 in Sioux City, IA on July 19, 1989. Dave Boxum, a KTIV - NBC cameraman was at the Sioux City airport at the time of the crash and captured video of the fiery accident. When he got back to the station, the local NBC affiliate, technicians tried to schedule transfer of the video to NBC in New York through NBC's Skypath system. The network was eager to get their hands on this exclusive coverage of the accident. When Skypath could not schedule the transfer immediately, KTIV gave the video to CONUS to forward to NBC New York via their transponder on K-2. Of course, as soon as CONUS got the video to New York they made it available to all of their other customers. Bye bye NBC exclusive! I always thought that some NBC heads had to roll over that one! CONUS was a subsidiary of Hubbard Broadcasting, St. Paul, MI. Stanley Hubbard was a pioneer in radio and television broadcasting and later was one of the principles in the formation of Direct-TV. When I visited CONUS, I always enjoyed looking at the pictures on the walls of Hubbard Broadcasting showing the early days of radio and TV broadcasting.

A typical Ku Band VSAT terminal consisting of an dish antenna with offset feed, and transmit/receive electronics in the box with cooling fins.   A coxial cable runs fron the electronics box to a modem inside the building. - *Astute Systems Photo*

There were also several Ku-Band transponders filled with Single Channel Per Carrier (SCPC) customers.   Most of these were data carriers for business customers. These systems used Very Small Aperture Terminal (VSAT) systems consisting of a small Ku-Band antenna, a built-in receiver and transmitter and a satellite modem.   Two K-2 transponders were filled with AT&T's customers.

I had written earlier that Americom had made a conscious business decision not to go into the the Very Small Aperture Terminal (VSAT) business.   The primary reason for that decision was the difficulties inherent in operating and maintaining a national network of small, unmanned stations located all over the United States -- and that decision was correct.   Now that Americom was part of GE, however, GE Plastics wanted a VSAT data network to interconnect their various manufacturing plants.   GE Plastics headquarters was in Pittsfield, Massachusets and manufacturing plants were in such places as Parkersburg, West Virginia, Selkirk, New York, Mt. Vernon, Indiana and Waterford and Fort Edwards, New York among a total of about 10 locations.   GE Plastics management had contacted Americom Marketing and asked them to set up a network to interconnect the computer facilities at each of the plants and their mainframe computers in Pittsfield, MA.   Americom management felt that they could not turn

down another division of GE, so suddenly we were in the VSAT business.   Americom
marketing and our engineering group hurriedly designed a system, bought hardware and
installed it at Pittsfield and the various plastics plants.   Unfortunately, they also did it on the
cheap.

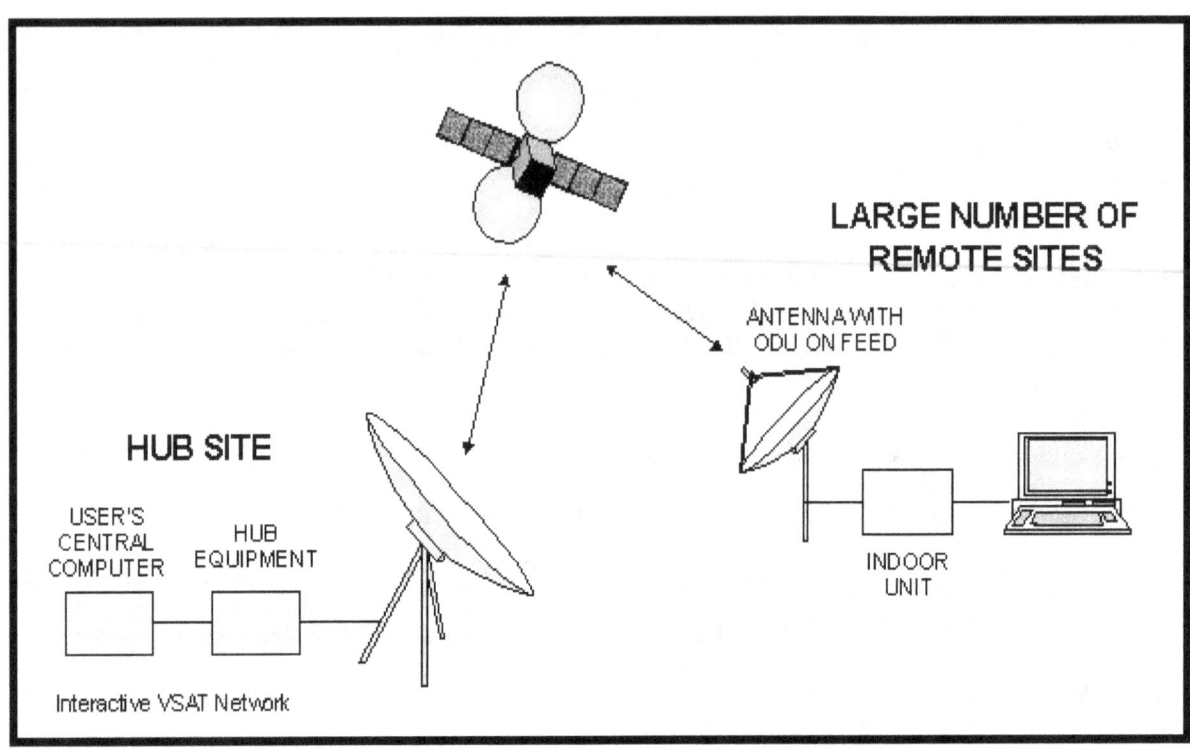

A VSAT Network showing the hub station and a typical remote station.   Because the satellite
beam covers the entire country, remote VSAT's can be located anywhere in the US. -
Diagram by macrosat.com

VSAT networks consist of a more robust hub station with more transmit power and capable of
higher data rates than the small VSAT terminals. All of the terminals consist of an antenna, an
RF package containing the transmitter and receiver and a data modem.   Although I had heard
rumors about what was going on with GE Plastics, no one had talked to Operations about the
network -- until they got into trouble!   About noon one day, I got a phone call from my boss
Carl Capista asking me if I could get to Pittsfield, Massachusets by that evening.   I said that I
probably could but asked what was going on.   Evidently the network had not been working too
well and the GE Plastics customer in Pittsfield was furious about the poor performance.   The
plan was that I was to meet the Americom Marketing and Engineering guys for dinner at a

motel in Pittsfield to be briefed on the system and its problems. The following morning, a meeting had been scheduled with the GE Plastics Information Technology (IT) Manager and the three of us. I was supposed to go to the meeting, make nice, explain how I was going to fix everything and smooth over the whole mess.

When I got to the motel, there was no sign of my Americom colleagues but a package of technical documentation had been FedExed to my room. As I read through it, I was appalled. There obviously had been little thought given to reliability and maintainability issues. My compatriots finally arrived and over dinner we discussed the problems. My first question was "Where is the backup transmitter for the Pittsfield hub site?" "It's in the building in the computer room" answered the engineer. "Why isn't it in the equipment box at the base of the antenna and wired so that it automatically switches online in event of a failure of the prime transmitter?" I asked. "There wasn't enough room in the box." was the answer. The hub and each of the VSAT terminals were set up with different data rates. I asked why each terminal had a fixed data rate modem when the modem manufacturer offered an agile modem that could be set to any required data rate -- one size fits all! No answer. The discussion went downhill from there.

The next morning I was introduced to the GE Plastics IT Manager who took us on a tour of the communication facilities. GE called the large main frame computers "Databases" and at Pittsfield they consisted of about a half-dozen large blue boxes with IBM logos on their side in a room in the basement. They provided the computing power to process all the production and sales data coming in from the plants all over the country over our VSAT network (hopefully). The IT guy had evidently been taking a lot of heat because the VSAT network had not demonstrated sufficient reliability to allow them to abandon their expensive terrestrial data communications network. When we looked at the hub installation, I tried to figure how to fit the backup transmitter in the fiberglass box-- it didn't look too hopeful. I looked at the deicing for the hub antenna -- it consisted of a 1000 Watt ladies hair dryer mounted inside the

243

antenna shroud! I made a note to get a couple spares because I doubted if they were very reliable -- I was right we replaced two that winter.

I told our GE customer that I would send a couple of techs up from Vernon to check out the hub. I also told him that we would buy a couple of the universal agile modems to give us more flexibility in case of failures. Our biggest problem though was that we didn't have a nationwide organization to respond to failures. At that moment we were dispatching techs from the nearest Americom earth station which was often an air flight away. We told him that we were actively looking for an organization that could provide nationwide support. After a reasonably cordial lunch we headed for home.

I sent the techs to Pittsfield and told them to check everything out and also to make sure that we had spares on hand for all the replaceable units.   In the meantime, Facilities Engineering reported that they were in discussions with NCR about handling the VSAT maintenance nationwide. NCR was the company that started out as National Cash Register and had grown into a large computer centered business.   Because they supported such vital equipment as point-of-sale terminals in supermarkets they already had a 24/7 maintenance organization in place.   They also had experience with satellite equipment because they had supported the Comsat Video Enterprises hotel pay-per-view TV project.   When I asked where NCR's office was in New Jersey -- the answer was "we have nine offices in New Jersey".   I came away from our first meeting hugely impressed.   Their proposal to support the Americom VSAT network included developing maintenance documentation for their techs and training the techs at their Dayton, Ohio training facility.   They showed us a clear 24/7 management escalation plan in case there were problems with support. Best of all they would do all of this for a fixed price of $600 per VSAT Terminal per year. NCR was as good as their word -- their performance for us was outstanding.

Now that we had maintenance support, I had to solve the spare parts problem.   After some digging, I found out that we could rent shelf space for our spares in FedEx's hub facility in

Memphis. I set that up and had a VSAT RF Package and a Universal Agile Modem stored on the Americom rented shelf at FedEx headquarters in Memphis. We researched each VSAT site to determine an address that would accept FedEx deliveries 24-hours a day. FedEx promised that if we called them before midnight, they would deliver parts from our shelf anywhere in the US by 10AM! In practice, when the NCR tech determined what spares he needed he would call the Americom Network Operations Center (NOC) 800 number and the Vernon Valley Americom tech would dispatch the parts from FedEx. The system worked great and pretty much solved our problems with our GE Plastics and other VSAT customers. Oh yes there were other customers! As soon as it was found out that GE Plastics had put us in the VSAT business, we had other requests for service.

The post mortem to this story is sad. We thought that NCR was a great company. They provided us with timely technical services for systems that were important to our business success. The technicians were excellent and went out of their way to get their jobs done. Most of the VSAT Terminals were on roofs that in the Northeast US were covered with snow for much of the winter. NCR would send out two techs to these sites so one could hold the other so they wouldn't slide off the roof. And then in 1991 NCR was bought by AT&T in a hostile takeover and in five years AT&T destroyed NCR and sold off its remnants. AT&T paid $7.4 billion for NCR and it was only worth $4.0 billion when they spun it off in 1996. NCR was a great company and deserved better than that.

Earlier I had mentioned that K-3 was our ground spare -- now that K-1 and K-2 were successfully launched and operating, there was less need for a spare. K-3 was to play a part in starting up the Luxembourg company that many years later would take over Americom. Actually the association with Luxembourg goes back to Americom's early days. Radio Luxembourg was a highly succesful enterprise that started broadcasting in 1933. They were really the forerunners of pirate radio because they broadcast to England in English and solicited advertising through their agents in the UK thus subverting the BBC's no advertising policy.

They also broadcast to France in French which severely irritated the French who also liked to control radio broadcasting. Radio Luxembourg made a lot of money.

In the late 70's we would get visitors to Vernon Valley from Radio Luxembourg on an almost annual schedule ostensibly because they were considering satellite programming in the US. In hindsight, I am certain that their real interest was in how the Americom satellite communications model would transfer to Europe. In 1983, a group of investors led by Clay Whitehead, head of the Office of Telecommunication Policy in the Nixon White House, initiated a project called Coronet to provide broadcast satellite services from Luxembourg to Europe. The project became a European political football and finally died in the face of strong French and Belgian opposition. Luxembourg, however, still wanted to develop a European satellite system. When they started to seriously consider formation of Société Européenne des Satellites (SES), Jacques Santer, the Prime Minister[11] of Luxembourg visited Americom facilities at Princeton and Vernon Valley and we had very interesting discussions with him. The Prime Minister had an excellent grasp of the business and technical issues and asked good questions. [12]

As you may remember, our buy from Astro was for K-1 and K-2 to be launched and a ground spare K-3. When SES got authorization to launch satellites, they were in a hurry to get on air and Americom President Kevin Sharer offered them K-3 which Astro had in storage in Delaware (no inventory taxes). They quickly accepted. Our K-3 was launched on an Ariane 44LPV-V27 launch vehicle from Kouru, French Guiana on December 11, 1988 and became Astra 1A. When Crimson Satellite Associates, Americom's partnership with HBO, fell apart one of the Crimson birds was sold to SES and became Astra 1B. Sharer racked up a lot of Concorde miles negotiating those deals. Both birds were collocated at 19 degrees East. I used to see SES people at Spacecraft Power Conferences because they had the same battery

---

11 Santer's official title was President of the Government, however, he was usually referred to as Prime Minister. In 1989 the title was officially changed to Prime Minister.
12 When the Prime Minister visited Vernon Valley, the Vernon Township police almost had apoplexy figuring out how their tiny force would protect a head of state. They had a couple of plain clothes guys follow us from the heliport but they lost us on the way back.

problems with Astra 1A that we had with K-1 and K-2. Astra 1B had some kind of thruster problems that caused them to dump traffic occasionally. In 1994, SES, now the largest communication satellite operator in the world bought Americom from GE and it became SES Americom. In 2010, SES dropped the Americom brand name and the company was folded into SES New Skies. RIP Americom.

I guess this would be as good a time as any to talk about the ill fated Crimson Satellite Associates project that came up about the same time as the Americom Ku Project. In fact Andy Hospodor, Americom President and CEO, and Bob Zitter, VP Operations for HBO, jointly announced the RCA-HBO joint venture during the launch activities for K-1 at Cape Canaveral. Crimson was a partnership between RCA Americom and HBO to supply Ku band services to cable operators and directly to homeowners (DTH). HBO thought that Ku Band was the next big thing. Remember, RCA had agreed to partner with HBO in Crimson as part of the deal releasing HBO from their contract for four transponders on F-4. The plan was for Americom to act as design agent to build two medium power Ku Band Astro spacecraft. HBO would supply free antennas to the cable operators to get them to convert to Ku Band and would market the project. The DTH service would be sold by the cable operators to homeowners in their area who were not passed by their cable system.

The two proposed spacecraft had sixteen 45 Watt, 54Mhz transponders. They had a 4850 Watt solar array and four 50AH batteries. The most interesting feature was that they had two 50-Newton, bi-propellant thrusters on the base plate instead of a solid motor AKM. With that arrangement the spacecraft is brought to synchronous altitude by a series of burns over several days rather than one enormous burn of the solid motor. I attended all of the design reviews at the Astro plant in East Windsor. I knew this was a little different venture because the HBO technical people always brought two lawyers with them to the design reviews. Fortunately the attorneys kept their peace, however, it made me nervous about our partners.

Artist concept of the Crimson Associates satellite. Unlike other Ku band spacecraft, it only had a dish antenna on one side. Bob Youngblood referred to it as "Van Gogh Sat". - *RCA Astro-Electronics Division Photo*

Unfortunately (or fortunately) the project fell apart about two years before the spacecraft were

to be launched. The cable operators had a huge investment in C-Band receiving systems and weren't the least bit interested in Ku-Band even with free antennas.   Also, it was a hard (if not impossible) sell to competitors like Viacom (Showtime) to convince them to put their business on an HBO owned satellite.   The big cable system operators campaigned strongly against any cable networks being distributed on a DTH satellite. After the deal fell apart, Americom was able to broker the sale of the two satellites.   One was sold to SES, renamed Astra-1B and was flown co-located with Astra 1A to provide service to Europe.   The other was sold to Intelsat and became Intelsat K and was located at 21.5 degrees W Lon. for Atlantic Basin services. Intelsat extensively reworked the antennas and switching to meet their unique international communication requirements. Unfortunately, Ed Horowitz who I always felt was one of the good guys at HBO, lost his job over the Crimson debacle.

The FCC had been issuing permits and trying to encourage Direct Broadcast Satellite (DBS) systems as competitors to Cable Television.   By 1989 none of the original eight applicants had been able to get financing or programming sufficient to start operations.   The FCC cancelled all of the original permits and established a new group of DBS applicants.   The FCC also realized that the DBS systems (like Crimson) were never going to succeed unless they had access to cable programming networks. In late 1992 Congress passed the Cable Television and Consumer Protection Act that guaranteed DBS companies access to cable programming networks and forbid cable television programming networks from discriminating against DBS companies.   These two events and technical advances particularly in digital video compression finally gave promise of success for DBS enterprises.   Out of this group of applicants through mergers and takeovers two operations emerged that remain today -- Echostar (DISH) and DirecTV.

After GE took over Americom, I went to many of the satellite trade shows. I remember one particularly well. It must have been the National Cable Television Association convention in Las Vegas in 1991.   This guy came into the GE exhibit and started talking to me and several of the marketing people about DBS.   He said that his plans were to launch six spacecraft to cover the US with direct to home broadcasting. I knew Charlie Ergen as a guy who had a fairly successful satellite dish business (Echosphere Corp.) that sold mainly to the backyard dish market. It was an interesting conversation and Charlie and I got into some of the details of spacecraft operations. When he left, however, I rolled my eyes at our marketing guys about Charlie's grand plans. To a man, those guys knew Charlie, and said "don't laugh, this guy thinks big". Charlie certainly did think, and act, big. In 1994, after I was retired and was consulting for AT&T, I ran into Charlie again. By this time, he was starting to implement his DISH network. He had made arrangements to buy two Series 7000 spacecraft from Astro. I was supervising operations of AT&T's Telstar-401, the first Series 7000 to be launched. We had an interesting discussion about the 7000 and how Charlie was going to use his two spacecraft. I still wondered whether he was going to be able to pull it off -- and of course he did. As of May 2011 EchoStar has launched 16 satellites, currently operates 14 and DISH Network has over 14 Million customers. Echostar operates an uplink facility near Cheyenne, WY.

The EchoStar uplink facility near Cheyenne, Wyoming. Many of these antennas downlink programming material from other satellites that is then uplinked to one of the EchoStar satellites for the DISH Network. - *Echostar Photo*

The other major DBS network is DirecTV a partnership of Hughes and USSB. Utilizing the leadership of Eddie Hartenstein (Hughes) and Stanley Hubbard (USSB), Hughes established DirecTV and launched its service on June 17, 1994. Hughes Electronics (A GMC subsidiary) launched 3 high powered satellites into the 101West DBS slot. To ensure its availability in rural areas, DirecTV signed an exclusive agreement with the National Rural Telecommunications Cooperative (NRTC) that allowed NRTC affiliates the right to market and distribute DirecTV in rural markets. In 2000, under pressure from its shareholders, GM authorized Hughes to seek buyers for DirecTV. In 2001 there were unsuccessful efforts made to sell DirecTV to News Corp. and EchoStar. The EchoStar deal failed on antitrust objections by the FCC and the Department of Justice. In the wake of that failure, EchoStar had to pay GM $600-million. In 2003 News Corp agreed to buy controlling interest in DirecTV subject to SEC approval. The SEC approved the deal with the proviso that News Corp agree to arbitration for all disputes with carriers of its media broadcasters and to provide content through DirecTV neutrally rather than favoring its own networks. DirecTV was the first and has been the most succesful DBS service with over 19 million subscribers.

DirecTV West Coast facility at El Segundo, CA. DBS broadcasters use large Ku Band dishes so that they can transmit more uplink power to the spacecraft to overcome rain fades. - *Antennas for Communications Photo*

Chapter 19 - Bigger Troubles

By 1985, F-3R and F-4 were approaching their half-lives. They were both completely filled with cable TV traffic and bringing in huge amounts of revenue. Among other customers, F-3R carried HBO East & West, Cinemax East & West, Showtime East & West, Nickelodeon, WTBS, ESPN, CNN & Headline News, Black Entertainment and The Weather Channel. Among others, F-4 carried Spanish International Network, Financial News Network, TNT, Bravo, Playboy, Home Sports Entertainment East & West and ABC and NBC Network Television feeds.   These were among Americom's the most valuable and important cable customers.

We had been figuratively holding our breath for years because both F-3R and F4 had serious hardware problems that had the potential to cause the loss of the spacecraft.   During F-3R checkout, right after launch, testing showed that Command Receiver 1 (CMR-1) had reduced sensitivity -- literally "hard of hearing".   It worked but required much higher than normal transmitter power levels from the ground to successfully execute commands.   As a result, we had been using CMR-2 ever since and keeping our fingers crossed that it would not fail. Occasionally we would test CMR-1 and the results were not encouraging -- its sensitivity appeared to be gradually deteriorating.

F-4 Telemetry Beacon 1 (BCN-1) had failed early in life and we were operating on BCN-2 -- again with our other fingers crossed.   If BCN-2 failed, we would receive no telemetry data from F-4 and would be literally blind -- we would have no idea what was going on in the spacecraft.

When F-3R CMR-2 failed, we knew that we were in deep trouble.   Over the years we had tested CMR-1 occasionally and knew that it was continuing to degrade -- it's "hearing" was getting worse. The klystron command transmitters in Vernon Valley and South Mountain had a maximum power output of 3000 Watts. Under normal conditions we had the transmitters set for about 200 Watts and had no problem commanding all of the spacecraft through either 10-meter, 12- meter or 13-meter ground system uplink antennas. In fact we had successfully commanded with less than 10-watts using a 13-meter antenna. Larger antennas have higher gain and thus transmit more power to the spacecraft. For that reason we dedicated the Datron 13-meter TT&C antenna to command F-3R. Even so, it was taking over 2000 Watts to command F-3R through this 13-meter antenna. We realized that we were going to require more power if we were going to be able to continue operating F-3R. We put a second command transmitter alongside the first and ganged their outputs together with a "Magic-Tee". Now we could get about 5000 Watts into the antenna. We also put transmitter controls and a remote power meter at the TT&C Console so that the Spacecraft Controller could adjust the transmit power level and read the uplink power. Every hour the controller sent a No-Operation (NO-OP) command list to F-3R at low power and gradually increased power until the commands verified in the spacecraft. He then logged the time, the CMR-1 temperature (from telemetry) and the transmit power level. We started plotting this threshold data and found out that the CMR-1 sensitivity varied quite a bit over the day and that variation appeared to be a function of temperature -- the warmer it was, the better it worked. We knew that as seasons changed CMR-1 was going to get colder than it was now. We added two more 3000 Watt transmitters to the two we already had -- now we had the "Gang of Four" and could put about 10,000 Watts into the uplink. This was about as much power as the antenna hardware could take --if we needed more uplink power we would have to find a bigger antenna.

I had an early Macintosh computer at that time and a spreadsheet program that could produce three dimensional graphs. I plotted the threshold data with time of day on the X-axis and CMR-1 threshold level on the Y-axis. So a single plot showed how the threshold varied over a 24-hour period. The three-dimensional graph arrayed the daily plots one behind the other with the most recent in the front. The effect was similar to a picture of a mountain range with the valleys being the best performance and the mountains the worst. It showed some interesting things. CMR-1 was mounted on the earth-facing panel right behind the slot between the two petals of the communication antenna. As sunlight came around as the spacecraft approached local midnight and the CMR was shadowed by the antenna petals, the CMR was cold and barely working. As the sun came further around and shined in the gap between the petals the CMR became quite warm and worked beautifully. The sun angle on the spacecraft changed daily. At the summer solstice there as full sun on the north face with none on the south face. As the fall equinox approaches the sun light moves south so that at equinox, it shines full on the side panels through the day with very little on the north and south faces. At the winter solstice, there is full sun on the south panel with none on the north panel. Obviously at all of the interim days the amount of sunlight varies on all the panels. Using the three dimensional plotted data, we tried to anticipate how the CMR would perform and plan evolutions requiring commanding accordingly. We also found out that the CMR sensitivity degraded as the Load Bus Voltage became lower such as during eclipses and when we had to turn the array off the sun during North/South maneuvers.

The lower bus voltage performance of the CMR bit us rather badly during an F-3R North South maneuver. During North/South maneuvers, the solar array had to be positioned bisecting the plumes from the north face thrusters. For most of the year this meant that the solar array was turned away from the sun and the spacecraft was running on batteries during the maneuver. Under those conditions the Load Bus Voltage was lower than when the array was on the sun. Because the CMR performance degraded as voltage went down, we reduced loads in the spacecraft so that the Load Bus would operate at the highest voltage possible. All redundant equipment and all possible heaters were turned off. The batteries were topped off right before the array was moved to insure that they were completely charged.

On this particular North/South maneuver the Solar Array was commanded into Reverse Slew to move it into position over the North Face thrusters. As it approached the desired position it was commanded to stop -- it did not stop!   Our command transmitters were at maximum power. We commanded again -- no joy.   I told the analyst to wait until the array was pointed at the sun and try again.   As the array passed the sun, the Load Bus Voltage increased but not up to normal levels. We repeatedly sent the Array Stop command with no luck. As the array again went off the sun, the Load Bus Voltage was even lower. As it came around to the sun again, the Load Bus Voltage went up but was lower than the last time on the sun because the batteries were discharging. We continued to try to command the array to stop every time it was pointed at the sun -- no luck.   By this time one of the analysts had determined that the problem was caused by a heater that was still on but should have been commanded off. This additional load had pulled the load bus voltage down to just below where the command receiver would work. Every time the array turned off the sun, the batteries discharged further and the load bus voltage dropped further. The spacecraft is designed to protect itself if the Load Bus Voltage becomes too low by turning transponders off. The protection logic turns off one transponder at a time, if the voltage drops to the danger level again it turns off another transponder and so on.   I was sure that as soon as the first transponder turned off, the Load Bus Voltage would come up far enough that the CMR would start working again.   I called the Network Operations Center (NOC) and told them what was going on.   I told them to monitor all of the television traffic on F-3R and as soon as a transponder went off to call the customer and tell them it would be back on in about five minutes.

About 10 minutes later, a transponder turned off, the Load Bus Voltage jumped up and we were able to command F-3R. As soon as the array was pointed at the sun we commanded the Solar Array Drive to Normal Forward and started recharging the batteries. We commanded the shut down transponder back on. The tech in the NOC called and said that he saw that the transponder was back on and that Black Entertainment Television (BET) was back on the air. I asked him if he had talked to someone at BET when the transponder went off. He said that he had talked to an engineer at BET and explained to him that we had a spacecraft problem and that the turn-off was not under our control.   By this time it was about 2AM.   The NOC tech

256

called again and said he had someone from BET on the line who wanted to talk to me. I told him to transfer the call and heard "This is Bob Johnson, President of Black Entertainment Television". In 1979, Robert F Johnson left the National Cable Television Association and with his wife Sheila, formed Black Entertainment Television (BET). BET had been quite successful and Johnson was often cited as the leading up and coming black business man in the US.[13] I apologized to him for taking BET off the air and explained that we had a spacecraft power problem and that the turn off was to protect the spacecraft and out of our control. Johnson was having none of it -- "You think that because we're a black company you can pull the plug on us and get away with it." I went through the whole scenario again in greater detail trying to explain that I had no control over which channel went off. Johnson refused to be mollified and kept repeating that it was just another case of discrimination. Finally after about a half hour of this, I told Johnson that what he was saying was absolutely not true and that I was sorry that he felt that way and terminated the call. I alerted my boss and the Video Marketing people that we had a problem with Bob Johnson, however, I never heard another word about it.

As the F-3R Command Receiver continued to degrade, I started to look around for a larger antenna that we might be able to use. I knew that the AT&T Hawley Earth Station had three 30-Meter antennas and had heard that they were not using one. I called George Johnson, the AT&T Spacecraft Operations Manager and he confirmed that the antenna was not in use. I immediately got in touch with the Engineering group in Princeton and asked them to see if we could lease their 30 meter antenna. A 30 meter antenna would give us a 7 dB advantage over our 13 Meter Datron antenna -- and we needed it.

---

13 Johnson ultimately took BET public and then sold it to Viacom for $3 Billion making him a billionaire and the richest black person in America until he was overtaken by Oprah.

Hawley AT&T Satellite Earth Station showing the three 30 meter antennas - AT&T Photo

A lease arrangement was quickly made and TT&C Engineering came up with a plan to install a computer in AT&T Hawley with one of Vernon Valley's Command and Range Tone Generators and connect it back to Vernon through a data link. Among other complications was the fact that software would have to be developed for the Hawley installation. I proposed a somewhat simpler solution. I pointed out that the command tones are all less than 20 kilohertz which is within the frequency range of commercial off-the-shelf systems used to distribute FM Radio programming by Ku Band satellite. If a 1-1/4" rack Transmit Unit was installed in Vernon and a 1-1/4" rack Receive Unit installed in Hawley, we could just transmit the command tones over our K-2 satellite radio link to Hawley and eliminate the computer and other hardware installation in Hawley. As usual with our TT&C Engineering group I had to fight the "Not Invented Here" syndrome but was able to prevail. Earth Station Engineering installed another "Gang of Four" transmitter arrangement in Hawley. They also installed a data link so that the Vernon Valley Controller could adjust and monitor the uplink power so that we could do threshold tests and we were in business. Because of the long command path, including a satellite radio link, our command verification software had to be adjusted slightly to compensate for the increased delays. The Hawley command system worked well and we were able to command F-3R in situations where we previously had failed. We still had times when we had trouble commanding and a number of rather hairy events, however, we were able to keep F-3R operating and bringing in large amounts of revenue for another five years.

I continued to worry about the fact that F-4 had only one telemetry beacon and if that one failed, we would be in deep trouble. The Satcom C-Band spacecraft contained two radio transmitter Telemetry Beacons -- one transmitting at 3.7 GHz in the horizontal polarization and one transmitting at 4.2 GHz in the vertical polarization. The beacon radio signals were received in the ground TT&C station and processed in the Data Computer. The beacons transmit telemetry data containing information on all of the spacecraft on-board equipment. Analog data includes electrical voltage and current levels, temperatures, attitude (roll, pitch, yaw) data and Solar Array Position. Discrete data includes equipment status (ON/OFF, HI/LO) and switch positions (A, B, C). Early in life, F-4 Telemetry Beacon-1 had failed. We could operate perfectly well with only Telemetry Beacon-2 (TLMBCN2), however, if it ever failed we would be operating completely blind. We would not get any data from the 225 points we monitored in telemetry and so would have no idea of what was going on in the spacecraft. We also could not see whether ground commands were received and verified in the spacecraft. Our ranging system used the telemetry beacon to send range tones back to the ground system -- without a beacon our ranging system was also dead.

At about 2 o'clock in a morning in November 1991, Spacecraft Controller Gene Lampkin called me and said "Arch, I think that the F-4 telemetry beacon just failed". When I got down to the station, I had Gene command the beacon off and back on again -- no joy. We tried again -- still no good. Then we tried Beacon 1 just in case it had come back to life -- dead! I told Gene to try turning the beacon on every hour in case the turn-off was temperature related.

259

Analysts John Bailey and Peter Staab were in the station by this time.   John was the Lead Analyst and Peter was responsible for F4. We got Bob Youngblood (who was now working in Princeton) on the speakerphone and discussed our options. I told them that we had to keep F-4 operating for as long as we could until Americom management decided how they were going to restore its customers.   We all agreed, that although it would be difficult, we could probably do East/West maneuvers and Momentum Adjusts in the blind.   The big problem, however, was to find some way to get range data so we would know where the spacecraft was.   I said that I would contact our marketing people and see if they could find an F-4 customer who would be willing to let us to put a small ranging carrier right alongside their TV carrier. Peter said that we had just done an F-4 North/South maneuver, had just finished ranging and gotten a new orbit. That was really good news for two reasons. We absolutely could not do a North/South maneuver in the blind and this would give us about a six week grace period before F-4 started to be out of its box. Also, we needed a good orbit to calibrate our new ranging carrier -- when and if we got one. I asked Peter to review his F-4 data and determine how long we had until we absolutely needed an East/West maneuver and a Momentum Adjust.   During the night I had briefed the two responsible Americom Vice Presidents on the situation and what we were working on. By about nine in the morning, I called them back and briefed them on our plans. The Operations VP, Rick Langhans said that he would work the ranging carrier problem with Marketing and our customers.

The need for Momentum Adjustments (MOMADJ) is a function of solar pressure and varies in frequency and direction throughout the year.   The solar pressure causes changes in the Momentum Wheel Assembly (MWA) speed.   If not corrected (unloaded) by firing thrusters, the wheel could become saturated (maximum speed) and the spacecraft would lose lock on the earth and start spinning around the pitch axis.   Fortunately, Peter had kept meticulous records over the past few years and had plotted out all of the MOMADJs.   I asked him to come up with a plan based on F-4's performance over this same time period last year. He could then come up with a MOMADJ schedule to maintain wheel speed in normal range for the next two months.   Unfortunately, at this time of the year when we were approaching the Winter Solstice, solar pressure was changing wheel speed quite a bit every day so we would probably require a number of adjustments.

Within a few hours Rick called and reported that Turner Network Television (TNT) had agreed to let us put a ranging carrier in the edge of their transponder.   Earth Station Engineering was doing the carrier power calculations and the Vernon Valley Earth Station Manager was rounding up the equipment required to put up the carrier.   We scheduled a test with TNT for the next day.   Time was of the essence because the further we got away from our last good orbit, the less accurate the calibration of the new ranging carrier would be.   The next day we prepared to bring up our ranging carrier at the edge of the TNT transponder.   We were on the phone with the TNT Chief Engineer and we both measured the Luminance Weighted Signal to Noise of the TNT video signal. We then slowly brought up the ranging carrier to the required power level calculated by engineering.   The TNT picture looked great -- we both took video noise measurements again and found that nothing had changed.   We then went through a ranging sequence where we modulated the carrier with a series of audio tones. The picture continued looking great and there was no change in measured video noise.   We thanked the Turner people and told them that the carrier would stay there indefinitely.   It took us about a day of test ranging and comparisons with our last orbit to get the system calibrated.   We then started on 30 hours of hourly ranging to see if we could get a definitive orbit that fitted with our last orbit.

Two things were really worrying me.   The first was what would we do if F-4 lost pitch lock on the earth.   With no telemetry, we wouldn't know which way the spacecraft went off the earth and thus would not know which thrusters to fire to get it back on the earth.   The second was that given the history of Command Receiver (CMR) failures on F-3R we could very well have the F-4 command receiver fail and not know it.   With no telemetry we could not get verification of commands sent to F-4 so we increased command transmitter power to over 1000

Watts -- the brute force approach.   I pondered these problems quite a bit.

I was in the shower one morning cogitating the F-4 pitch lock problem when I had one of those "Eureka" moments.   If we knew whether our New Jersey or California station lost our ranging carrier signal first, we could figure out which way the spacecraft was rotating.   When I got to the station, I had one of the techs connect the Ranging Receiver's Signal Level (AGC) voltage to a Strip Chart recorder pen.   I also had him put an output from our station clock on an adjacent pen.   We set up the same arrangement at South Mountain. Now we had a continuous plot that would tell us the exact time that each station lost the signal.   If Vernon lost the signal first, it would mean that F-4 was turning westward.   Conversely, if South Mountain lost the signal first it meant it was turning eastward.   Fortunately, we never had to use it -- Peter's planning and execution of MOMADJ's worked very well.

The command receiver problem was really bugging me.   Being a firm believer in Murphy's Law, I was certain that the Command Receiver would fail in the middle of something critical like an East/West maneuver.   I kept thinking that if there was something that we could command off and see the result, we could test the receiver.   What could we turn off -- why a transponder of course!   Right -- turn off a transponder with customer traffic in it.   And then it would take 2-1/2 minutes to come back on.   The more I thought about it though I thought -- well maybe.   I talked to Video Marketing in Princeton and said "Not all of our F-4 customers are major players. For instance, there are a couple of Mom and Pop shopping channels on F-4. I bet that they would let us turn their channel off for 3-minutes at a time of their choosing if you offered them some rebate dollars".   And sure enough they did. One of the small shopping channels, America's Value Channel in Transponder 5, agreed to our deal and I expected that they would want to do it in the middle of the night.   It turned out that they wanted to do it at exactly 5:10PM Eastern time.   I had no idea why and I didn't ask because it worked out great for us.   Once a day at 5:10 PM we started sending Transponder 5 OFF commands at very low power while watching the TV monitor -- we kept increasing power until we saw their picture disappear.   We logged the transmitter power level and commanded the transponder back on.   The tests did show that the command receiver did take a little more power than normal, however, it didn't change over time and that was reassuring.

While all this was going on Americom was looking for a communication spacecraft that they could lease. They found that Telesat Canada's Anik D2 had just been replaced with a new bird Anik-E2 and there was no longer any traffic in D2.   Americom closed the deal with Telesat, bought Anik D2[14] and Telesat started moving it along the equator from it's current location at 111 degrees West Longitude to F-4's location at 83 degrees West. The journey would take about four weeks.   Our challenge , in the interim, was to keep F-4 flying reliably until our customers could be seamlessly transferred to D2.   Our bigger challenge, after that, would be to push F-4 westward climbing out of synchronous altitude and keep it under control -- still with no telemetry.

Peter had come up with a plan for Momentum Adjustments (MOMDAJ) based on where MWA Wheel Speed was in the last telemetry we received before the beacon failed and how much wheel speed had changed last year at the same time.   The plan was reviewed by Bob Youngblood and by Astro's Orbital Dynamics Group.   We executed the MOMADJ thruster firings without incident.   Of course, with no telemetry we didn't know whether we got the desired correction.

A couple of days later, we did our first F-4 East/West maneuver.   First, I confirmed that we had the command transmitter set to maximum power output.   We had developed a special Maneuver Plan checklist for F4 East/Wests.   We sent the command lists to warm up the thrusters about a half-hour before the maneuver start time.   I called the Network Operations Center (NOC) and told them to monitor all of the F4 TV traffic and to notify us immediately if

---

14 In 1993 after Americom no longer needed Anik D2, they sold it to ArabSat who designated it as Arabsat-1D to backup the ailing Arabsat-1C Spacebus 100 spacecraft.

there was any degradation of any of the TV signals. We sent the command list to set up the spacecraft for an East/West maneuver followed by the start command list. It was an eery feeling -- we had sent all of the commands in the blind and we had no idea whether they had been received or whether they were executing. As soon as the scheduled end of the maneuver passed we started ranging. We also had Vernon and South Mountain take Tracking Antenna azimuth and elevation data every half hour and we plotted it to see if F-4's drift had changed -- it looked like it had. We could only range from Vernon so we ranged every half hour instead of every hour. It would take at least 24 hours until we had enough data to obtain a definitive orbit, however, after a few hours Peter ran an orbit with the few ranges that we had. Although not a great orbit, it showed that the semi-major axis had changed in the direction that we expected -- Halleluah! By late the next day we had a definitive orbit that showed that we had gotten the exact drift and eccentricity correction that we had planned. We successfully executed another East/West and a couple of small Momentum Adjusts as Anik-D2 approached from the west.

Telesat Canada Spacecraft Operations had started Anik-D2 moving eastward over four weeks ago. Now it was drifting slowly toward F-4 with it's Communication Receivers off. When a synchronous spacecraft is moving eastward, it is below synchronous altitude; when moving westward it is above synchronous. As Anik-D2 approached, we had maneuvered F-4 so that it was at the eastern edge of the 83 degree West "box". As Anik-D2 entered the western edge of the box, we executed an East/West maneuver (in the blind) on F-4 to start it moving westward. As F-4 passed above Anik-D2 in approximately the center of the box on December 6, 1991, Anik-D2's communications receivers were turned on and then F-4's were commanded off -- and we got command verification when we saw the F-4 video signals disappear. The transition from F-4 to Anik-D2 was made without serious interruption to traffic. All of the cable operators had been told that the Anik-D2 polarization angle was different and that they would have to adjust their antennas. As F-4 continued westward, Telesat Canada executed an East/West maneuver to stop D2 in the 83 West box.

Our last problem with F-4 was to get it going fast enough westward so that it would be above synchronous altitude and out of the way of other synchronous satellites. We were following F-4 with the Vernon Valley and the South Mountain 13-meter TT&C Antennas. George Johnson, AT&T's Spacecraft Operations Manager had offered the use of the their TT&C Antenna at Hawley, PA. to help us track F-4. We could only range through the Vernon antenna but azimuth and elevation data from the other two antennas would help us more accurately track F-4s exact location. We started doing fairly large East/West maneuvers to accelerate F-4 west. We kept increasing the size of the thruster pulses. The azimuth and elevation data confirmed that F-4 had picked up speed. We were ranging every half hour and running orbit determinations. All the data showed that our thruster burns were successful and we were driving F-4 westward. As we increased the pulse lengths, we could see the normally straight line trace on the ranging carrier signal strength strip chart recording start to become sinusoidal. This was caused by the spacecraft's nutating (coning) motion because of the large burns and was to be expected -- another confirmation of success. After one burn, the nutation increased quite a bit. Bob Youngblood said that we were probably running out of fuel and one of the thruster sets (odd or even) was blowing nitrogen pressurant instead of firing. He suggested that we open the cross-connect valve between the odd and even fuel systems and equalize the fuel in the two systems. Bob had often opined that if you equalized when there was a big pressure difference between the odd and even tanks, the resultant fuel surge would cause a large spacecraft disturbance. We commanded the cross-connect valve open and watched the strip chart recorder. Well there must have been quite a surge because all of a sudden there was hardly any nutation in the spacecraft. The fuel surge must have occurred at a nutation peak and effectively "denuted" the spacecraft -- we could not have timed it better if we tried, We did a few more big pulses and decided to call it quits. All of our tracking data showed that F-4 was well above synchronous altitude heading westward. We commanded all systems off -- R.I.P F-4.

I sent out a letter of congratulation to all those that had been involved. We had done something that, to my knowledge, no one had done before -- successfully operated a revenue bearing communication spacecraft on station for six weeks with absolutely no telemetry data -- and then disposed of it in accordance with International Agreements. F-4 was a very successful spacecraft -- designed for 10-years, it probably earned RCA about $300 million in its 9-year lifetime. I was very proud of my people and all of those who helped us -- they rose to the occasion and did a great job. The night that the second beacon failed there was much doom and gloom and cries of "What will we do now?" That night I told everybody "look, this is a hero opportunity -- let's buckle down and fly this thing!" And they did -- and they were heroes.

A chronic problem with all of the Satcom spacecraft was that, once they were on-station in space, they were much warmer than the pre-launch predicted temperatures. Some of this was probably caused by a miscalculation of the alpha factor in the design of the Optical Solar Radiators (OSR). OSR's have a high emissivity that allows them to reject heat into space while at the same time having a low absorbancy of the heat from the sun. The resulting device looks exactly like a mirror. The glass mirror surface of the OSR can become contaminated for a number of reasons and that contamination reduces the OSR emissivity and thus reduces how much heat it can radiate into space. The body of the spacecraft, which is pelted with radiation, tends to accumulate an electrostatic charge. This charge attracts particles and when they land on the OSR they further reduce its reflectivity and emissivity. But where in space do these particles come from? Several places it turns out. When a spacecraft that has been constructed in the earth's atmosphere goes into space there is quite a bit of outgassing as the outside pressure reduces from 15 psi to zero.. The reduction in pressure causes the various material in the spacecraft to outgas chemical compounds which then find their way to the surface of the OSR's. Also when the Apogee Kick Motor is fired to put the spacecraft into synchronous orbit is fired it ejects many materials, some of which, also find their way to the OSR's. Over the life of the spacecraft, hydrazine fueled thrusters are fired during maneuvers and also deposit materials on the OSR's. It is the thruster contaminants that most likely cause the continued degradation of the OSR's and the increase in spacecraft temperatures over lifetime.

Shortly after launch, we had observed that K-1 and K-2 were quite a bit warmer than we expected. This was not too big a surprise because all of the Astro birds had been warmer than predicted. As time went on we realized that the K-birds were much warmer and the batteries in particular were hot. Nickel-hydrogen (NiH2) batteries have a unique combination of advantages of energy density, cycle life and reliability -- but they don't like to be hot. The optimum temperature range for NiH2 batteries is between 10-degrees and 15-degrees C. For every degree C outside that range, particularly on the high side, the battery loses 1 Ampere-Hour of capacity. We did not realize that at first, because we did not have too much experience with the new NiH2 batteries. Our previous experience with Nickel-Cadmium (NiCad) batteries was that although we operated them outside the designed temperature range, it didn't seem to bother them much if at all.

Nickel-Hydrogen batteries consist of 40 or 50 individual cells wired in series to obtain the desired operating voltage. Each cell has a voltage of 1.5 volts when under load. The electrode elements of each cell are encased in a stainless steel cylinder pressurized to 1200 pounds per square inch (PSI) with hydrogen gas. The primary characteristic of Nickel Hydrogen batteries is long life.

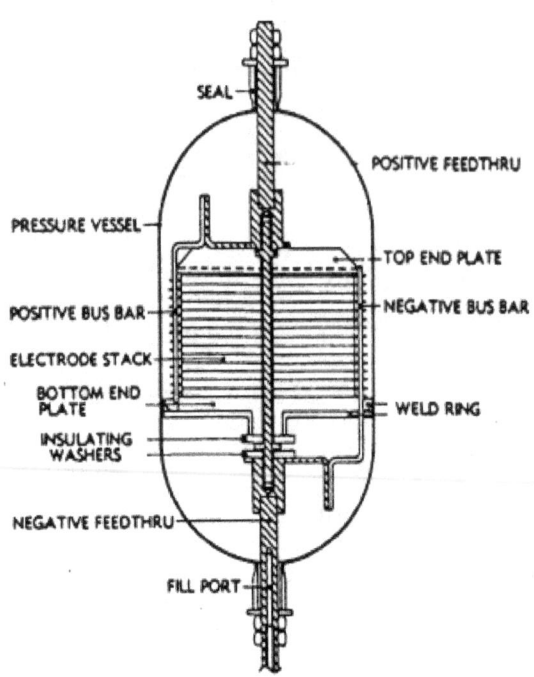

Typical Nickel-Hydrogen cell designed for spacecraft use. The cells are about 3-inches in diameter and about 10-inches long. When combined in a battery, the cells are usually arranged in a "wine rack" configuration - *NASA Diagram*

During the first eclipse season in the Spring of 1986 we didn't have power problems because K1 and K2 were still pretty lightly loaded so the TWTA's were not drawing excessive current. By September of 1986, as we entered eclipse season, K2 has heavily loaded including NBC's eight transponders filled with saturated (maximum power) video carriers. Nickel-Hydrogen batteries are not supposed to have the "memory"effect as NiCads do and so do not require deep discharge reconditioning before eclipse season. Because we had little experience with them, I had elected however to recondition the batteries in both K1 and K2 in August. Autumnal eclipse season starts at about September 1st with a few minutes of eclipse around spacecraft midnight. The eclipse length increases each night until the Autumnal Equinox on about September 21 when the eclipse is 72-minutes long. As the batteries discharge during eclipse, they become endothermic and get hot. When they are being recharged, they are exothermic and get cold.

As we got into the second week of eclipse season, Jack Schmidt my power specialist came to me and expressed some anxiety about the K2 batteries. He said that they were getting very hot (over 30 degrees) and their discharge current during eclipse was higher than he anticipated. The K-2 load bus voltage was dropping to abnormally low levels and he was concerned that the bus could drop to the transponder shut-off point.[15] The power system is designed so that if the Load Bus Voltage gets too low, the spacecraft protective system starts to turn off transponders one at a time to save the spacecraft. Jack had discussed the situation with the power specialist in our engineering group in Princeton and they had come up with some non-critical loads that could be turned off during eclipse. I told him to turn those off and to put the batteries in High Charge about an hour before eclipse to make sure that they were "topped off" but not over heated.

As we got to the night of Equinox, the Load Bus Voltage dropped far enough that one of NBC's transponders shut off. It was about 1:20AM in Vernon, however, the transponder that turned off was carrying the NBC West Coast traffic where it was 10:20PM -- prime time! And then it hit the fan. NBC Skypath was on the phone to the NOC wanting to know what happened. The Spacecraft Controller followed procedure and tried to command the transponder back on but it wouldn't stay on because the Load Bus was so low. We still had about a half-hour until K2 came out of eclipse. I called Skypath and told them that we had a serious problem and that the transponder wouldn't be back until we got out of eclipse in about 30 minutes. Fortunately NBC had a backup West Coast feed on our F1R C-Band bird and Skypath had their West Coast affiliates switch to that feed. As soon as K2 came out of eclipse we were able to command the transponder back on but the damage had been done. Skypath reported to the NBC management that we had a serious problem with K2. First thing in the morning there was a conference call between NBC management and the Americom Marketing and Engineering personnel. NBC was reassured that the K2 spacecraft itself was well and in no imminent danger of failing. The problem that we were having with batteries was correctable by changes

---

15 The K-Band power subsystem was designed based on the power required for 42-Watt TWTA's. Later, 45-Watt TWTA's became available and Americom elected to use them based on the margins available in the Power Subsystem. As a result we had less power margins available when we started to have battery problems.

in battery charging procedure, however, it may take some time. In the interim NBC decided to use their C-Band West Coast feed on our F1R spacecraft as primary until the problem was resolved. I wished that I was as confident as our management that the problem was solvable.

My spacecraft analysts and the Americom Engineering Group had been looking at alternate charging plans that made use of the spacecraft daily temperature curves to try to keep the batteries cooler. The normal plan for recharging spacecraft batteries during eclipse season was to command the batteries in to High Charge just as eclipse starts. As the batteries supply all spacecraft power during the eclipse and discharge they get hot. The longer the eclipse, the hotter they get. As soon as the spacecraft comes out of eclipse and there is sufficient sun on the solar array, the battery charger starts recharging at the highest charge rate. The batteries start to cool down as the charging current causes an endothermic chemical reaction in the battery cells. After a while as the battery becomes fully charged the battery voltage increases and the temperature rises slightly. At that point the battery is fully charged and the charger is switched to trickle charge which is just enough current to keep the battery topped off without causing overheating. This normal procedure wasn't working for K-2.

Every day as the spacecraft goes around the earth, the sun angle changes. At spacecraft dawn the sun is on the east face, at noon on the back panel, at dusk on the west panel and at midnight on the earth facing antenna panel. As a result, the temperatures of the components in the spacecraft vary during the day depending on their location in the spacecraft and over the year as the north/south sun angle changes. The three batteries were located in different parts of the spacecraft so their temperatures varied independently. In consultation with the Americom and Astro battery experts, we set up separate charging plan for each of the three batteries. Basically the plan was not to start charging after coming out of eclipse. Instead we waited until the battery was past its hottest part of the day and was starting to cool down. The battery was then commanded into High Charge, became endothermic and got even cooler. We carefully monitored the Battery State of Charge, Battery Voltage and temperature. When our calculations showed that the battery was 30-minutes away from being fully charged, we commanded it into Trickle Charge to just maintain the charge it had. Thirty minutes before eclipse, we commanded the battery back into High Charge and it started to get cooler. If our calculations were correct, the battery would be fully charged and just starting to go exothermic and get warmer as the spacecraft entered eclipse. Because the length of eclipse changed from day to day, all of the parameters of the charging algorithm had to be recalculated daily -- for three batteries each for two spacecraft! There wasn't much room for error -- if you didn't get it right you could either have the bird going into eclipse with partially charged batteries or have them overcharged and hot. Jack Schmidt spent many hours analyzing the previous night's battery performance data recalculating all the parameters and times and generating Operations Bulletins to direct the Spacecraft Controllers on the times that all commands were to be sent. It was a huge amount of critical work.

We survived the first eclipse season, however, it was very clear that we had a serious battery problem. Whatever was going on in the cells when they got hot had caused a loss of battery capacity. The three batteries, when new, each had 50 ampere-hour capacity. Our calculations showed that they now had substantially less than that. We had a similar problem with Nickel-Cadmium (NiCad) batteries in older spacecraft. With NiCads, we reconditioned them by deep discharging down to a low voltage and the capacity was restored. Deep discharge didn't fix the Nickel-Hydrogen batteries. Americom management asked the GE Corporate Staff if there were any resources within the corporation that might be able to help. A group in GE Laboratories in Schenectady agreed to look at our problem. They were not battery people but were very experienced in dealing with chemical processes in closed pressure vessels -- which is what our batteries were.

What happens to batteries that lose capacity is that some of the chemicals get to a state or a place where they are no longer available to the charge/discharge process. Deep discharging NiCad batteries allows the cadmium crystals that collected on the terminal to return to the basic cadmium electrolyte.   Deep discharge followed by recharge didn't help the Nickel-Hydrogen batteries.   The people at GE Labs ran a number of tests and experiments on Nickel-Hydrogen cells and measured the cell capacity. They found out that if a battery is completely discharged, left in that state for several weeks and then recharged, the normal chemical state is restored. Although this was a time consuming process, it restored the batteries and allowed almost normal operations.   The batteries were still hotter than they should be and required special charging procedures and we had to monitor their performance very closely -- but we could carry all the communication loads through eclipse.

K-1 and K-2 were very succesful birds and made Americom a lot of money.   K-1 was in service for 11 years and K-2 was in service for 17 years.

Chapter 20 - "Their hearts really aren't in it"

One big change that occurred after Americom became part of GE is that we had very productive management meetings and they were held in nice places. One that I remembered particularly was in Key Biscayne, Florida in January 1987. The Winter Management Meetings always began on Super Bowl Sunday. We would all arrive in the early afternoon and then there would be a buffet dinner and drinks while we watched the Super Bowl. Super Bowl XXI had the New York Giants beating the Denver Broncos 17 to 2. Being a Giants fan, I thought that it was a great game. Monday and Tuesday were non-stop meetings. Wednesday was meetings until noon and the the afternoon was for recreation. Some played golf and some went deep sea fishing. I opted for snorkeling at John Pennycamp Coral Reef State Park down in the keys. As the name implies, the park is all underwater on the reefs offshore from Key Largo.

It was a fairly long drive from Key Biscayne to Key Largo so immediately after the end of the morning meetings, we were all loaded into a bus with box lunches and a cooler of drinks. It was an interesting group that included Kevin Sharer, Americom President, Neil Bauer, Exec. VP Marketing, Walter Braun, Exec. VP Engineering and Operations, John Christopher, VP Technical Operations, Rick Langhans, VP Engineering and a number of other technical and marketing people. As we finished our lunches, the conversation turned to AT&T. Sharer had worked with AT&T and knew most of the principles in Skynet, their satellite group. He said that AT&T was only in the satellite communication business because as the preeminent US communications company they felt that they had to be in it. "Their hearts really aren't in it" he said. He said that many people within AT&T felt that Skynet was outside of their core business interests. Sharer thought that AT&T would be open to proposal for either outright purchase of Skynet or for some kind of a joint venture between AT&T and Americom. That opened up about a half an hour of spirited discussion with various proposals for takeovers or joint ventures being kicked around. As the discussion went on and beers were consumed everyone seemed to think that this was a great idea and that we could make a deal with AT&T.

We finally got to Key Largo and had a nice afternoon of snorkeling and sunbathing. I thought that the AT&T discussion had been a fun exercise and nothing more than that -- I didn't expect anything to come of it.

About two weeks later I got a call from Rick Langhans. He told me that Sharer had talked to Karl Savatiel, AT&T's Director of Satellite Communications, and that AT&T was interested in a possible business arrangement between Americom and Skynet. Interested enough, in fact, that there was a meeting at AT&T in Bedminster, NJ on the next day at 10AM -- could I make it? Of course I could! The AT&T facility in Bedminster was a 200 acre complex with about 70,000 square feet of office and technical space. They must have had a lot of folks working there because I had a helluva time finding a place to park. The Americom team assembled in the building lobby. Besides myself and Rick there was Carl Capista from Program Management and a couple of marketing guys. After we were greeted and badged we were given a quick tour of the AT&T Network Operations Center.

AT&T Global Network Operations Center Bedminster, NJ - *AT&T Photo*

After looking at the span of AT&T's global operations displayed in their operations center, I could understand why AT&T might feel that the US Domestic satellite communications business was just a side show to their core businesses.

We were ushered to a large conference room upstairs and met the AT&T team. George Johnson, who was AT&T's Spacecraft Operations Manager, and I were old friends -- I didn't know any of the other people. AT&T definitely did not want anyone to get wind of what was going on. We were escorted through the building and they brought lunch to us in the conference room instead of us going to the cafeteria. We discussed in very general terms how our two operations could be combined without ever talking about who would be in charge. We agreed that, initially, we each needed to familiarize ourselves with the others business. Rick, Carl and I and a marketing guy made arrangements to visit an AT&T earth station at Coram Long Island. A phony story was concocted and agreed upon to explain why GE people were visiting AT&T facilities and vice versa.

This was all pretty new territory to the AT&T people -- they had just recently come out into the competitive world. In 1982, AT&T and the Justice Department agreed on terms for settlement of an anti-trust suit filed against AT&T in 1974. AT&T agreed to divest itself of its local telephone operations (New York Telephone, Pacific Telephone, etc.). On January 1, 1984 the Bell System ceased to exist. In its place were seven Regional Bell Operating Companies (RBOC's) and a new AT&T that retained its long distance telephone, manufacturing, and research and development operations. The RBOC's were Ameritech, Bell Atlantic, Bell South, NYNEX, Pacific Telesis, Southwestern Bell and US West. Over the years, through mergers and takeovers the RBOC's were reduced to three -- reorganization, splits and mergers continue to this day. The primary reason that AT&T agreed to the breakup was to get permission to enter the computer business. AT&T Data Systems Group turned out to be an ill fated venture that only lasted about 10-years and destroyed NCR, a great computer company, in the process (see Chapter 16).

At the time of our story, AT&T was 3-years out of the breakup and still in the process of converting from a tariffed, regulated common carrier operating under the Communications Act of 1934 to a competitive free market corporation.   Before the breakup, AT&T calculated costs for each of their services (in fact for individual paths of those services) plus a profit and submitted proposed tariffs to the FCC for approval.   Because there was no competition, there was no incentive to reduce costs so facilities were designed and built on a "Cadillac" model. [16] Bell Labs had plenty of money to research and design things and Western Electric built them to the highest standards.   As we talked with AT&T it became obvious that they were just then reorienting to a system where they could determine what the true costs of their operations were. As one of the AT&T guys said to us "if a facility carried 10% of the traffic then it got assigned 10% of the costs".   They were trying to establish all of the cost factors involved in the business and assign them to the appropriate facilities or business units.   At one point, we were comparing my Vernon Valley station and AT&T's Hawley station which were about equivalent size.   One of us asked AT&T what it cost to run Hawley each year.   They gave us a number and I looked at him and said my electric bill is more than that -- they had a way to go.

Carl and I went out to the Coram Long Island earth station.   It had a single 32-Meter antenna pointed at the Intelsat bird at 1-degree West Longitude.   The earth station was a revelation to us. Never had we seen a station constructed as if cost were no object.   Everything from the racks to the cable trays were custom designed -- probably by Bell Labs.   We talked to the station manager about how they implemented new services.   Any reconfiguration of the station went through their Plant Engineering Department and could take anywhere from six months to a year to turn up new services.   Carl and I were appalled -- we were used to getting new customers up on the bird in weeks not months.

A few weeks later. we visited the AT&T earth station at Hawley, PA that served customers in

---

16 For instance, round "tom-tom" antennas are used in most microwave networks. Sometimes a round horn antenna is used if necessary to reduce interference. By contrast, every one of AT&T's microwave links used large Bell Labs-designed square horn antennas that also required a massive tower to support the weight.

the New York Philadelphia area. The station had three 30-Meter antennas with cryogenic parametric preamplifiers operating at -16 degrees Kelvin. This was serious (and expensive) overkill. GE and the other domestic satellite carriers used 9 or 10 Meter antennas with thermoelectric parametric amplifiers operating at -35 degrees Kelvin and carried traffic equal to or greater than AT&T. In 1985 right before Americom went out of the PLC telephone business we were carrying over 3000 telephone circuits in our Single Side-Band (SSB) satellite carriers. By comparison, AT&T was carrying 1200 telephone circuits in their FM/FDM satellite carriers. While most earth stations had diesel generators to supply backup power in the event of commercial power failure, Hawley had three Solar Gas Turbine Generators. The whole station was over designed. Part of the reason for that lies in the history of how AT&T went into the US domestic satellite communications business.

In the early 1970's when RCA and Western Union were each hard at work building satellites and ground systems to establish their own domestic satellite communications systems, AT&T was taking a much more cautious approach. They asked RCA Astro-Electronics, Comsat General and Hughes Space Systems for proposals to build a complete turnkey satellite system consisting of 3 spacecraft and seven terrestrial earth stations located throughout the US. The succesful bidder would also launch and operate the spacecraft and lease back the communications capacity to AT&T. AT&T seemed to be very worried about possible problems they might have operating communication spacecraft. At the bidders conference for this project, the AT&T presenter was asked why they wanted a lease arrangement. The answer was "we don't want a Three Mile Island".

Comsat General Corp., the domestic operations subsidiary of Comsat was the succesful bidder. Comsat had built all of the US earth stations for the Intelsat system. Comsat General proposed to build Intelsat Standard A Earth Stations for the seven AT&T terrestrial facilities. "A" Stations were designed to receive the relatively weak signals from Intelsat spacecraft global beams and thus required a 30 Meter antenna. I never understood why they sold AT&T

these huge antennas for a US Domestic Satellite System where the received signals were 6 to 7 dB stronger than Intelsat.   At Americom, we considered that for most applications an 11 Meter dish would work well -- and they did.   We hedged our bets a bit by installing 13 Meter antennas in our first earth stations -- we definitely considered 30 Meter dishes[17] serious overkill.

Comsat General had Hughes build three HS-351, 24 transponder, C-Band spacecraft for the AT&T Project.   Comstar 1A was launched from Cape Canaveral on an Atlas SLV-3D on May 13, 1976 and placed into geosynchronous orbit at 128-degrees West Longitude.   Comstar 1B followed on July 22, 1976 and was placed at 95 West.   Comstar 1C was launched two years later on June 29, 1978 and stationed at 87 West.   The three spacecraft were operated by Comsat General from their TT&C facility in Clarksburg, MD.   The spacecraft were initially lightly loaded primarily because AT&T was constrained from fully entering the US Domestic Satellite business until January 1979.   In fact, when Americom lost F-3 on launch in December 1979, they leased Comstar D2 (nee Comstar 1B) from AT&T because it wasn't in use.

---

17 Of course it is sometimes handy to have a 30-Meter dish in the neighborhood -- like when the F-3R command receiver was failing and we borrowed AT&T's 30-Meter antenna to get more power into the bird!

AT&T Hawley Earth Station with its three 30-Meter antennas. For comparison note the 13-Meter TT&C antenna. - *AT&T Photo*

As the Comstar spacecraft aged, AT&T decided to buy the next generation spacecraft and operate them themselves. In 1980 they contracted with Hughes Space Systems for three Hughes 376, 24 transponder, C-Band spacecraft. The contract also included TT&C spacecraft control facilities to be installed at the AT&T earth stations at Hawley, PA and at Three Peaks Petaluma, CA. Hughes also installed 13-Meter TT&C antennas at Hawley and Petaluma. Telstar 301 was launched from Cape Canaveral on July 28, 1983 on a Delta 3920/PAM and stationed at 96 West. Telstar 302 was launched on Space Shuttle 41D on September 1, 1984 and stationed at 86 West. Telstar 303 was launched on Space Shuttle 51G on June 19, 1985 and stationed at 86 West.

The meetings and negotiations between AT&T and Americom continued on. I built an enormous Excel spreadsheet with the Americom cost data on the left, the AT&T cost data on the right and a spreadsheet that calculated the differences in the middle. After about a year, the AT&T people began to get a handle on what the real numbers were and their data improved. At the same time, we were looking at how a combined venture might operate.

I came up with a plan that showed that Americom could operate all of its spacecraft and all of AT&T's Hughes spacecraft from a combined facility at Vernon Valley. In 1985 we had moved the Vernon Valley TT&C facility into the first floor of a new building. The second floor of the building was completely vacant -- we certainly had the room to expand. As part of the upgrades to the TT&C hardware and software system, we had gone to a database driven telemetry and command system. Processing the Hughes PCM telemetry data would primarily involve building new telemetry databases and a modest amount of software modifications. The command side was a somewhat more difficult because of the precisely timed commands required to fire thrusters on a spacecraft spinning at 60 RPM. All in all, it could be done. The bigger problem was the RF path to the AT&T satellites. Because of a land use dispute with Vernon Township, GE was under a moratorium on installation of new antennas. The Hawley Earth Station's link to the outside world was a microwave radio link from Hawley to the AT&T Colesville Microwave facility near High Point New Jersey. Americom's Hamburg Mountain microwave facility could see Colesville from its tower so we could establish a microwave link from Vernon Valley to Hawley to interconnect to the Vernon TT&C to Hawley's uplinks and downlinks.

It was a great plan but alas it never came to fruition. After three years of meetings and negotiation a plan for a Joint Venture between AT&T Skynet and GE Americom was agreed on. After approval internally in both companies, the plan was submitted to GE and AT&T outside counsel for legal opinions. Unfortunately, the verdict from both parties counsel was unanimous -- the deal would never pass Federal antitrust scrutiny.

I believe that Kevin Sharer was right -- AT&T never seemed to take the satellite business seriously. The AT&T people that I was dealing with were technical people who had a vested interest in the satellite business and their own jobs. Over time, however, I got the feeling that AT&T would like a way to ease out of most of the business but still keep a finger in it because they were AT&T. In 1994, after I was retired, I consulted for both AT&T and Martin Marietta on the Telstar-4 program. I spent a lot of time at Hawley and from what I could see AT&T management still didn't have their arms around Skynet. They were very disorganized. There were six supervisors in the Hawley Earth Station reporting to six different people in Bedminster and no one was actually in charge of Hawley. In 1997 AT&T sold Skynet to Loral Space and Communications. AT&T's Vice President of Product Management Business Markets, Bob Aquilina said "AT&T's strategy is to take advantage of the many opportunities opened by the new telecommunications regulation by focusing on our core businesses and customer franchises. While Skynet is a strong business and is expected to grow, selling it makes sense for AT&T because it is not central to our new strategy." Sharer was right -- their heart was never in it. Also in 1997, Loral bought controlling interest in Telesat Canada and merged Skynet into it. Loral's satellite communication services, including the Skynet Telestar satellites, are now marketed under the name "Telesat".

Chapter 21 Epilogue

It had always been my intention to work until I was 65. As 1991 came to a close, Americom was being reorganized again and I didn't feel like going through the inevitable agita. I would be 62, the minimum qualifying age for Social Security, on March 10, 1992 -- it was time. I put in my papers and retired from GE Americom on May 1, 1992.

My 18 years with Americom was the best job that I ever had. It was certainly challenging from the first day when I realized that the F-1 launch was only six months away and we were far from ready. It was exciting to be part of an enterprise that was building a business using cutting edge space technology to achieve an economic advantage in a very competitive marketplace. On some days it was exciting in a different way as we discovered some of the complexities and surprises of operating first-of-a-kind spacecraft. I derived a great deal of satisfaction in driving the continually evolving ever improving process of Spacecraft Operations. We had to change endlessly to accommodate the requirements for new spacecraft, work around spacecraft failures and, more importantly, to incorporate lessons learned from experience -- it was a process of continuous change. We certainly made our share of mistakes -- but hopefully learned from each one.

What made it all easier was the support we had from Americom Spacecraft Engineering and from the Astro engineers and scientists. Some of the problems that we had were complex and not easy to solve, but they worked through them with us until we had an operationally effective solution. It was very comforting to know that help was only a phone call away. One of my greatest frustrations, however, was TT&C Engineering. They didn't understand what we were doing, didn't listen and were owners of the "not invented here syndrome". It was too bad because I depended on them for the TT&C hardware and software and it seemed like it always took a confrontation to get what we needed to do our job.

Being one of a very small group of people experienced in Communications Spacecraft Operations, I did consulting for seven years after I retired. I did work for GE, AT&T, Martin Marietta and Lockheed Martin besides some other smaller less well known companies. I spent almost a year providing support to AT&T and Lockheed Martin for Telstar-401 the first Astro Series 7000 spacecraft. Catherine and I lived in Hong Kong during 1995 and 1996 when I was providing support for AsiaSat-2 an Astro Series 7000 Communications Spacecraft for Asia Satellite Telecommunications (AsiaSat). In 1998 we lived in Beijing while I was supporting launch and early orbit operations of an advanced Astro A2100 Communications Spacecraft, ChinaStar-1, for ChinaStar Telecommunications Satellite Co., Ltd. It was a great experience, which for the most part I enjoyed greatly.

The domestic communication satellite business has changed radically since Americom started operations in 1975. In those days, Americoms only real competition was AT&T and Western Union -- and each of us was only flying a few spacecraft. As time went on other companies went into the business each with only a few spacecraft. Over the years a consolidation of Fixed Satellite Service (FSS) operators took place. (See figure below) Some failed (Westar and SBS), some bought out others (Hughes and Contel-ASC) or were bought out (ASC and PanAmSat). Today as this is being written there are really only three FSS operators in the North American market -- SES World Skies with 26 spacecraft, Intelsat with 51 spacecraft and Telesat with 13 spacecraft. In the beginning, Americom and the other US carriers were limited by US Government regulation to providing services only to the United States -- the reason that Americom was spun of from Globcom. Thanks to the hard work and lawsuits by Rene Anselmo of PanAmSat, Intelsat's monopoly on international satellite communications was broken and now almost all FSS operators provide domestic and international services.

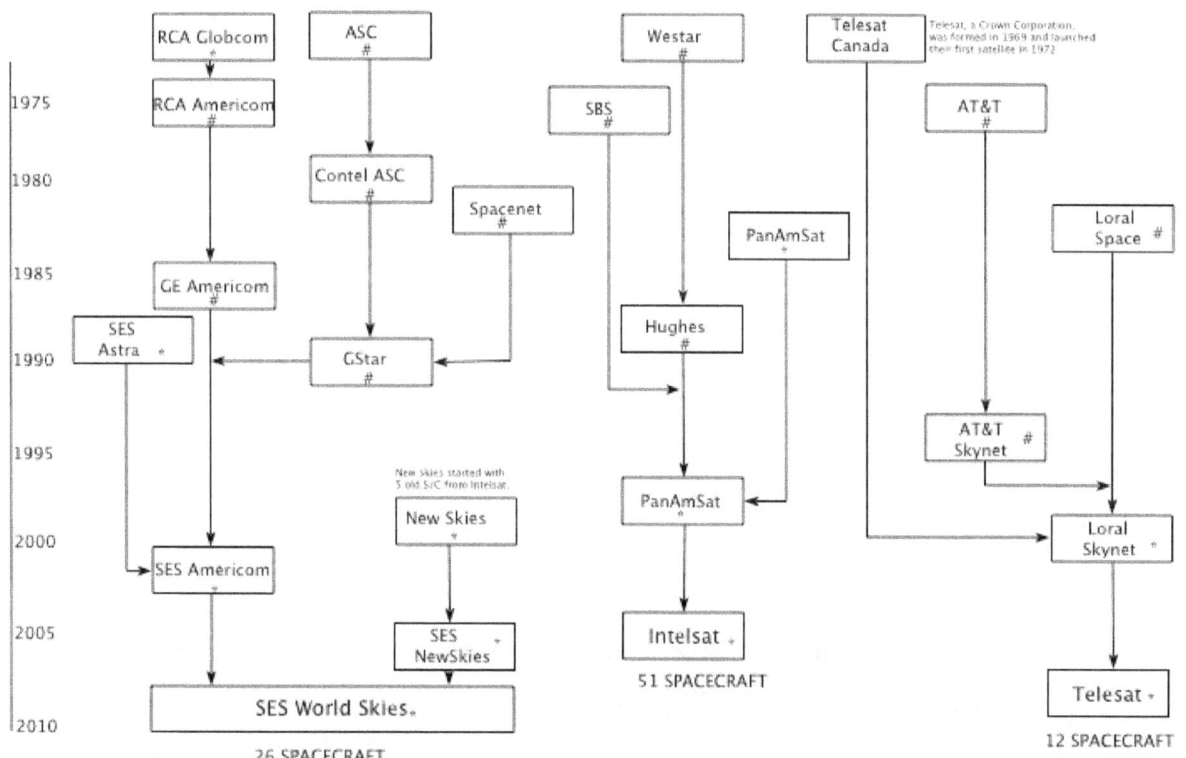

EVOLUTION OF NORTH AMERICAN FSS SATELLITE COMMUNICATION COMPANIES – 1973 TO 2011

ATM – 7/17/2011

Walter H. Braun
Senior Vice President-Government and
Technical Operations

GE American Communications, Inc.
Four Research Way, Princeton, NJ 08540-6684
609.987.4172,3

April 10, 1992

Dear Archie,

Only a commitment made months ago to attend the Goddard Dinner in Washington DC this evening prevents me from being present to pay tribute to you personally.

Archie, you have been a singularly important and key contributor to Americom since the very beginning of this organization. I feel compelled to praise and acknowledge your expertise, your commitment to quality and your dedication to having the finest possible operations department.

I will confess that being your supervisor occasionally made me feel like I was riding a bucking bronco in a rodeo and hanging on for dear life, but I say that with utmost regard and warmth.

You were a powerful advocate for your position, which is as it should be when a person has your responsibilities.

You'll have completed 34 years with us when you retire, but you are an energetic and resourceful person. I know that you will be participating in and contributing to other organizations as you go forward in life, and I wish you smooth sailing, clear skies and the best of fortune in everything you do.

Archie, with respect, affection and a touch of sadness, we bid you adieu.